<u>Praise for *The Frankenstein Fix*</u>

"An outstanding and impressive book. Technology has invaded our habits of life and our modes of thinking without our approval—even without our awareness. How did we get to this dangerous place in so short a time? And what can we do about it? In this sweeping but concise history of the digital age, Gabriel Cohen aims to show us the freedoms we've unconsciously surrendered, and points to ways to achieve a more reasonable future, in which seeds of destruction aren't planted in the business model. I learned a hundred things that I didn't understand or appreciate that all happened when I thought I was paying attention."

—Tom Zoellner, Ph.D., winner of the National Book Critics Circle Award for Nonfiction for *Island on Fire*

"I really like *The Frankenstein Fix* for the way it provides clear, actionable alternatives to tech doom."

—Douglas Rushkoff, author of *Throwing Rocks at the Google Bus* and *Team Human*

"This book really gets at the roots of why the culture and mentality of too many of today's tech leaders can lead to harmful consequences for all of us."

—Katy Cook, psychologist and author of *The Psychology of Silicon Valley*

"Cohen kicks off his central argument with a fascinating foray through the accidents of history whereby inventors created sweeping socio-cultural changes through unintended "side effects" of their technological advancements. Next, he walks us down a disquieting memory lane of digital firsts: The first custom-targeted internet advertisement! The O.G. SPAM message! The dawn of omnipresent AI! It's haunting to see this progression laid out on the page, step-by-invasive-step, effectively explaining the back-story behind tech companies' profiteering through extraction of our personal data and unpaid labor. Fortunately, Cohen doesn't leave us stuck in hopeless acquiescence to Big Tech. Instead, he shares actionable strategies for how to push back against technologists who prioritize "surveillance capitalism" over human health and safety. Grounded in history and an unflinching analysis that's unafraid to point fingers, Cohen's book engages the most critical questions of our age."

—Jessica DuLong, CNN columnist and author of *Saved at The Seawall: Stories from The September 11 Boat Lift*

"*The Frankenstein Fix* is more than a dire warning; it's also a blueprint for action. As technology takes us beyond science into a realm of ballooning egos and ever-higher profits, a future without restraints, ethics, or remorse, it falls on us to supply those vital qualities. Gabriel Cohen shows us how."

—Wayne Grady, author of *Technology*

"*The Frankenstein Fix* is a sharp and impassioned plea for a better tech industry and a better world. Cohen shows us how a lethal combination of lack of empathy and unintended consequences got us to where we are, and provides some compelling suggestions for how we can chart a different course. Recommended reading for anyone who believes we deserve a better future than the one currently being peddled by Silicon Valley."

—Wendy Liu, author of *Abolish Silicon Valley: How to Liberate Technology from Capitalism*

"Gabriel Cohen, a master of spinning mysteries, now unravels his biggest one: why the negative unintended consequences of new tech are often greater than the positive intended ones. He tracks down the clues, interrogates the subject, and pronounces his verdict. An absorbing, real-world whodunnit—and howwecandoit differently."

—Andrew Boyd, author of *I Want A Better Catastrophe: Navigating the Climate Crisis with Grief, Hope, and Gallows Humor*

<u>Other Books by Gabriel Cohen</u>

[An] outstanding first novel."

—*New York Times Book Review* on *Red Hook*

"Intricate, atmospheric, funny, and enthralling . . . an impressive crime novel from a powerful, promising new writer."

—George Pelecanos on *The Graving Dock*

"This elegantly written novel tackles wrenching questions in an unflinching, kindhearted way.

— *Time Out New York* (four-star review) on *Boombox*

"Encouraging and accessible throughout, Cohen's book will make a useful tool for readers going through a difficult break-up."

—*Publishers Weekly* on *Storms Can't Hurt the Sky: A Buddhist Path Through Divorce*

An impeccable procedural plus a poignant love story, intelligent, understated, and refreshing."

—*Kirkus Reviews* (starred review) on *Neptune Avenue*

"Spellbinding...Deftly plotted and convincingly written."

—*Kirkus Reviews* (starred review) on *The Ninth Step*

The Frankenstein Fix

Why Big Tech Goes Astray & What We Can Do About It

Gabriel Cohen

E ❖ A ❖ P

EDGEWOOD
AVENUE
PRESS

Book Cover Design by ebooklaunch.com

ISBN 979-8-9934624-0-0 (paperback)

ISBN 979-8-9934624-1-7 (ebook)

Identifiers: BISAC: SOCIAL SCIENCE/ Technology Studies; COMPUTERS/ Social Aspects; TECHNOLOGY & ENGINEERING/ Social Aspects

Edgewood Avenue Press

New York

"FIX"

1: *a position of difficulty or embarrassment; predicament*

2: *repair, mend, restore, cure*[1]

Contents

Introduction: On Sharks & Killer Coconuts   1

Part One
PROBLEMS

1. "This Changed Everything": The Advent of
   Internet Advertising   17
2. Clicks And Consequences: Psychological And
   Social Effects   30
3. Cozy Communities And Creepy Clusters:
   Political Effects   42
4. What If Different Choices Had Been Made?   64
5. "The Sword And The Spear": The Danger Of
   Technological Arms Races   72

Part Two
CAUSES

6. "The Technical Problem Captivated Me": On
   Psychology And Culture   87
7. On History, And Progress   107
8. How We Think Of Tech   123
9. Up to the Stars: Space Colonization and
   Messaging E.T.s   133
10. On Ideology: A Very Strange Icon   149
11. The Elephants In The Room: Venture Capitalists,
    Monopolists, And Hyper-Growth   163

Part Three
SOLUTIONS

12. Science Fiction: Sounds Of Thunder   183
13. A Predictor's Toolbox   190
14. Better Questions: At The Center Of The Wheel   211
15. A Shared Fight   219

16. System Repairs   242
17. A Cornucopia Of Other Solutions   261
18. Other Ways To Go   285
19. Rising To The Challenge   312

Acknowledgments   325
About the Author   327
Bibliography   329
Notes   333

Introduction: On Sharks
& Killer Coconuts

Here's a fact that never ceases to blow my mind.

In a 2023 survey about artificial intelligence, several thousand participants were asked, "What probability do you put on human inability to control future advanced A.I. systems causing human extinction or similarly permanent and severe disempowerment of the human species?"[1]

The average response was ten percent.

What's astounding about this was that the respondents were not anti-tech critics—these were people who actually worked in the field.[a]

A ten percent chance is not a negligible one. Historian Yuval Harari and technologists Aza Raskin and Tristan Harris put this in a helpful perspective when they suggested imagining that "as you are boarding an airplane, half the engineers who built it tell you

---

[a] In that same year, a number of noted pioneers of the A.I. field put this probability of calamity—which has become known as *(p)doom*—even higher. (https://www.fastcompany.com/90994526/pdoom-explained-how-to-calculate-your-score-on-ai-apocalypse-metric)

there is a ten percent chance the plane will crash, killing you and everyone else on it."[2]

They followed up with the obvious question: "Would you still board?"

I can't answer for you, but I certainly wouldn't. And I'd like to believe that if I started building something, but realized that it would present a substantial possibility of accidentally destroying not only myself, but also *the entire human race*, I would stop—or at least slow down considerably, until I could guarantee a safer product.

At the time of the survey, even though they acknowledged that terrible risk, a small number of tech moguls were eagerly forcing all of humanity onto the A.I. plane—whether we wanted to climb aboard or not.

They still are.

This isn't a book about whether or not the potential dangers of artificial intelligence will be realized. It's about two much broader questions.

What forces tend to push new technologies in harmful directions, in all sorts of fields?

And for a safer, healthier world, how can we push back?

This is also not a book about people who set out to intentionally cause harm.

The creators of technologies may be spurred by the purest of motivations: a passion for scientific discovery, a zeal for solving problems, an urge to create, or even a simple joy in play. They're proud of how their inventions can overcome the natural limitations of our bodies, enabling us to move faster, work stronger, see farther, speak louder, and heal better. Of course, some might just be looking to make gobs of money, but I think it's reasonable to suppose that most start out with benevolent aims.

Let's listen, for example, to how Silicon Valley investor Roger McNamee describes one company's beginnings in 2004. "From its earliest days," he writes, "Facebook was a company of people with good intentions. In the years I knew them best, the Facebook team focused on attracting the largest possible audience, not on monetization. Persuasive technology and manipulation never came up. It was all babies and puppies and sharing with friends."[3]

By October, 2021, however, when Facebook data engineer Frances Haugen blew the whistle on her company in public interviews and hearings, she lamented how she and her colleagues had unintentionally contributed to psychological distress in young girls, promoted "an information environment that is full of angry, hateful, polarizing content," and spurred ethnic and political violence around the world.[4]

What happened? Mark Zuckerberg and the founders of other Big Tech[b] platforms certainly didn't mean to contribute to any of those bad consequences. How did we get from their intended results to the ways in which so many things have gone so far astray? What powerful forces acted upon them? It's critical that we identify those forces, because the same factors that caused Facebook to go wrong are now at work on the developers of other nascent technologies, from bioengineering to autonomous weapons, from quantum computing to space colonization.

If we want to consider how things might go wrong with new

---

[b]   This term has sometimes been used to refer to a handful of the top tech companies in the U.S. (especially Alphabet, Amazon, Apple, Meta, and Microsoft), but I'll use it to refer to large tech companies in general.

technologies, it will be helpful to examine how things have already gone astray with others. For example, we can look at the history of the internet.

As a professor of college classes in journalism and critical thinking, I ask my students how many people they think get killed by sharks worldwide every year.

One tentatively raises her hand. "Five hundred?"

I shake my head.

Across the conference table, another student ventures a guess. "A thousand?"

I smile. "Not even close. On average, it's five or six."[5] I write that number on the whiteboard, then turn to face them again. "Okay, now how many sharks do you think get killed by humans?"

A student gives it a stab. "Ten thousand?"

I shake my head and point upward.

"Fifty thousand?"

I write a different figure. "Actually, it seems to be as many as a hundred million."[6]

The students' eyes widen.

"Some are killed for food, like shark fin soup," I tell them, "but many just get caught in nets for other fish." I cross my arms. "Now, who do you think should fear the other more, sharks or humans?"

I talk about the power of statistics to make or break an argument, and point out that I took care to ensure that my figures came from reputable sources. I found them while writing a magazine article[7] about a conflict between the Humane Society and the operators of a shark-fishing tournament. Some of the competitors argued that sharks deserve to die because they're a big threat to humans. When you look at the facts, that's a tough claim to defend.

I tell the class that while I was researching shark attacks, I kept encountering another interesting statistic. All over the internet, in all sorts of major news outlets, I read that more people get killed by falling coconuts—150 poor souls per year. When I read that grimly amusing factoid I was eager to include it in my article, but then I probed a bit deeper and discovered that it's entirely false. It seems that a doctor once wrote a medical article about somebody getting injured by a coconut, someone else exaggerated the story, and then a chain went on, like the childhood game of Telephone, until even otherwise-dependable journalists settled on that "definitive" number.[8]

I find the internet incredibly useful, and it has made my work easier by a remarkable degree. These days, however, I don't have to tell you that it has become the world's greatest source of _inaccurate_ material.

It certainly didn't start out that way. The people who invented it were academics in the U.S. who wanted to link computers at several universities[c] so they could share information. I doubt that any of them foresaw that their new network would soon be clogged with wild _mis_information (not to mention porn, spam, and savage user comments). After the 2016 U.S. presidential election and Great Britain's Brexit referendum, many observers seemed shocked to discover how useful the Web had proved for those who wanted to _intentionally_ spread false info, including wild conspiracy theories.[d] And the net-based platform Twitter gave a candidate in a democracy the ability to sidestep the press and communicate to his supporters directly and instantaneously, without any of the fact checking that might ordinarily accompany a politician's statements. As Donald Trump himself put it, during a rally in Phoenix, Arizona on August 22, 2017, "If I didn't have social media, I wouldn't be able to get the word out. I probably wouldn't be standing here, right?"[9]

---

[c] The first four universities connected were UCLA, Stanford, UC Santa Barbara, and the University of Utah.

[d] _Misinformation_ is unintentionally false or misleading; _disinformation_ is intentionally so.

These unexpected uses of a new technology should not have been so surprising. After all, there has been a clear tendency throughout human history: *Each time we introduce new technologies, we create them for some specific practical purpose, but often fail to predict their psychological, social, economic, and political ramifications. We think of these consequences as "side effects," yet they're often more far-ranging, powerful, and profound than the intended ones.*

The printing press provides one of the greatest historical examples. When Johannes Gutenberg invented his version of it in the 15[e] century,[e] he just wanted to make it easier to print multiple copies of pamphlets and books, but the machine had enormous social and political effects: its use has been credited with increasing literacy, spreading scientific data and methods, transforming education, and diminishing the power of monarchies. Media theorist Marshall McLuhan also linked it to the rise of nationalism, capitalism, and democracy.[10]

Another theorist, Neil Postman, also weighed in on the invention, and coming across an insight from him was one of the things that prompted me to write this book:

> "Every machine is an idea, or conglomeration of ideas. But they are not the sort of ideas that lead an inventor to conceive of a machine in the first place. We cannot know what was in Gutenberg's mind that led him to connect a winepress to book manufacturing, but it is a safe conjecture that he had no intention of amplifying individualism or, for that matter, of undermining the authority of the Catholic Church. There is a sense in which all inventors are, to use Arthur Koestler's word, sleepwalkers. Or we

---

[e] Although Gutenberg designed an excellent new press, Chinese inventors had developed printing techniques as early as 800 A.D., and used moveable type 200 years before him. (https://lithub.com/so-gutenberg-didnt-actually-invent-the-print ing-press/)

might call them Frankensteins, and the entire process the Frankenstein Syndrome: one creates a machine for a particular and limited purpose. But once the machine is built, we discover—sometimes to our horror, usually to our discomfort, always to our surprise—that it has ideas of its own; that it is quite capable not only of changing our habits but [...] our habits of mind."[11,f]

I was intrigued by this view of how unintended consequences of technologies can shake up our societies, and even transform our very selves.

At first, I thought I might just write a diverting history of inventions that brought unexpected results. Sometimes this happens due to simple accident. In 1856, for example, chemist William Perkin was trying to make quinine from coal tar, but he inadvertently synthesized mauve, the first artificial dye.[12] (This would turn out to be a little mistake with very big outward ripples: after Perkin's discovery, a number of German companies prospered by creating other artificial dyes, helping the nation transform itself from an agriculture-focused country into an industrial powerhouse—right before World War I.[13])

Other effects build in complex, inadvertent chains. Historian James Burke notes that the chemicals that saturated gun cotton, invented in 1846 as an alternative to gunpowder, enabled the development of a substitute for ivory after a shortage of billiard balls some twenty years later, and the shells of those balls were

---

f   McLuhan also argued that our mass consumption of printed matter fundamentally reshaped the way that human beings think, initiating a shift from thousands of years of oral culture to one based on vision.

made of what was named *celluloid*, which expedited the rise of the motion picture industry.[14]

It would have been fun to write a book about such outcomes, but it wouldn't have been very helpful. If unintended consequences are out of our control, we can't do much to avoid the bad ones.

As I did more research, however, I found increasing evidence of a fundamental truth: *Though bad consequences of technologies may be unintended, that doesn't necessarily make them random or accidental. In fact, we can identify a common set of factors—psychological, cultural, ideological, political, and economic—that make them likely to arise.*

I'm going to call this array of causes the *Wheel*.

Here's a related point: *Just because a consequence was not intended, that doesn't necessarily mean that it could not have been predicted.*

Of course, it can be difficult—and sometimes impossible—to anticipate technologies that might arise in centuries or even decades ahead. For example, as science writer Steven Johnson has pointed out, "Almost no one predicted the network-connected personal computer. Numerous science-fiction narratives—starting with H.G. Well's vision of a 'global brain'—imagined some kind of centralized, mechanical superintelligence that could be consulted for advice on humanity's biggest problems. But the idea of computers becoming effectively home appliances—cheap, portable, and employed for everyday tasks—seems to have been almost completely inconceivable."[15]

It might be tempting to throw up our hands and say, "Oh well, no one can see into the future." For our purposes here, however, when it comes to prediction we have one big advantage. We won't try to guess what new technologies might be invented in decades to come. Instead, we'll look at some that are already in development. We know what these technologies are; we just have to guess what they might *do*. And we have a bunch of interesting and effective forecasting techniques, which we'll explore in the pages ahead. They're often used to predict financial or military outcomes, but we'll consider how we might apply them to psychological, social, and political ones.

Again, if we want to do a better job of considering what might go wrong in the future, it will be helpful to examine what has gone wrong in the past.

To look at a recent example, could looking into history have helped the founders of Facebook, Twitter, and YouTube see big problems coming after they made their platforms increasingly effective at engaging users, and made it ever easier to spread information?

The answer would seem to be a resounding Yes.

For one thing, tech critics had already predicted a number of those problems. Roger McNamee, who was one of Mark Zuckerberg's mentors and early investors, tried to alert him to political abuses before the 2016 election,[16] but others had been warning about possible dangers of social media for decades. When I interviewed media theorist Douglas Rushkoff, who has been writing about the impacts of digital technology since the early days of the internet, he told me that as far back as 1994, "when *Wired* magazine announced that 'real estate online is infinite, but that human eyeball hours are fixed,' and told us that we're living in an 'attention economy,' many people realized, 'Oh my, if companies start competing for our attention using technology, bad things will happen'."[17]

Zuckerberg and the others might have also foreseen trouble if they looked further back. We might think of the internet as unprecedented, yet it had forebears in older communications technologies. Adolph Hitler made strong use of one of them. His government subsidized the production of millions of cheap radios and distributed them to German citizens,[18] and in 1933 he began giving speeches over the airwaves. Joseph Goebbels, his Minister of Propaganda, soon proclaimed that, "It would not have been possible for us to take power or to use it in the ways we have without the radio."[19,g]

As this precedent makes clear, enabling a greater range and flow of information does not necessarily bring positive consequences for societies.

Here's an even earlier example: we think of the internet as an information superhighway, so we might take that metaphor literally and consider the ancient Romans, who built a network of superbly engineered roads that enabled them to spread their empire far and wide. What they failed to take into account is that highways enable traffic in more than one direction—a fact that Germanic tribes from northern Europe exploited when they used those same excellent roads to tromp into the center of the network and sack the capital.[20]

Even though Mark Zuckerberg is a fan of the Roman emperor Augustus,[21] who helped develop those roads,[22] he seems to have

---

g   The Fuhrer also took advantage of the new, highly sensitive Neumann CMV3 microphone, and of improved public address systems. Notes microphone expert Klaus Heyne: "All of a sudden, Hitler was able to be more evocative, more animated in his delivery because [...] he could actually use the technology to help him deliver by being able to move, by being soft, by whispering and building up, and then getting super loud and it would still sound good. I bet you it was just the same as when the Beatles performed at Shea Stadium. All of a sudden, there was a dimension that was part of his delivery, and of the masses responding in the same way, uniformly, that was infectious. The microphone was able to allow that. It's a feedback, and it goes back and forth, and it gets to a crescendo. [...T]echnology, the microphone all the way to the loudspeaker, amplified the emotional translation of the message. I don't even know what that would be now. Is it Twitter, that [...] some people use very effectively?" (From *The Meaning of Hitler*, a 2020 documentary directed by Petra Emmerlein)

missed that particular lesson. In 2017, seemingly chastened by criticism of his company's role in creating all sorts of unintended problems, he wrote a manifesto in which he announced his company's new official mission, which was to "give people the power to build community and bring the world closer together."[23] That sounds warm and cuddly, until you realize that he was still only envisioning his platform's intended traffic, ignoring the fact that it provided equal passage for old high school pals and election deniers, for cat lovers and members of Al-Qaeda. And it didn't just provide the highway—Facebook's algorithms,[h] designed to boost user engagement, steered users toward wilder, more radical lanes.

As we go along in this book, we'll look at a number of ways in which the past has informed the present. And we'll examine how the current situation in Silicon Valley is heightening the potential for things to go wrong ahead.

I'll say this up front: we'll spend a lot more time looking at negative consequences of technologies than positive ones. That's not because I can't appreciate their many good aspects. As I write this, I'm sitting in an air-conditioned room under the bright light of a halogen lamp, wearing reading glasses and machine-woven clothes, typing on a stunningly powerful computer, using an astoundingly helpful internet search engine. Above all, I'm sitting here alive—though I was hit with a major health problem a few years ago, my doctors were able to resolve it by using highly sophisticated biomedical techniques.

It's clear that technologies make our lives better in a myriad of

---

[h] An algorithm is a sequence of instructions about how to perform a calculation or accomplish a task. (A kitchen recipe is an algorithm for cooking something.) Computer algorithms can cause operations to take place automatically, without human intervention.

ways.[i] Right now, though, it's especially urgent that we consider those that might lead to negative consequences. When we consider history, we often think of the power of monarchs or elected officials, but since the Industrial Revolution, much power to shape our lives has been captured by non-political agents, a relatively small number of technologists, who often release world-changing inventions without any public input or control. In the digital era they have shifted into particularly high gear, and there has never been a more crucial time for us to take stock of what they're creating.

Thankfully, in the past several years, revelations about threats to our mental health, our privacy, our jobs, and even our democracies seem to be making many people more aware of the potential dangers of new technologies. Though the first half of this book will detail all sorts of problems with them—and their creators—the second will be more upbeat, offering hope for a better future.

To ensure that this will not be just an outsider's critique, I have included testimony from many people who have worked inside Silicon Valley and other centers of high tech, and wherever possible, used their own voices to help tell the tale. I also interviewed experts in a wide range of fields, including a tech journalist, a historian of science, a psychologist, a digital media theorist, a tech philosopher, a financial tech executive and coder, a professor of international relations, and a forecasting specialist.

We live with countless forms of technology, ranging from construction machinery to medical devices to manufacturing tools, but I'll focus on those that rely on digital software, tend to be funded by venture capital, and are often referred to as "tech." In

---

[i] On the other hand, it's not easy to think of a technology that has had no adverse unintended effects. It's tempting to choose medical equipment and techniques—surely there's no downside there? Yet overuse of antibiotics has led to drug-resistant "superbugs," and technologies that make it possible to extend lives can also prolong suffering after any real quality of life is gone.

recognition of the fact that these are not just being created in California, when I refer to "Silicon Valley," I'll often be using it as an umbrella term for all the places around the world where tech is being developed today, from Beijing to London, from Bangalore to Tel Aviv.

Excellent books have been written on individual technologies mentioned here, but this one will take a broader approach, striving to identify root causes that engender trouble across a number of seemingly unrelated fields. Once we have identified those, we'll examine specific techniques that can help us do a better job of foreseeing—and avoiding—negative consequences. Then we'll look at how people in the past have managed to actively resist inventions that threatened their livelihoods and lives.

Finally, we'll consider how we might push back against the rash Doctor Frankensteins of Big Tech, in order to reduce dangers and make our world safer and healthier for everyone.

Part One
PROBLEMS

Before we get into discussing new inventions, let's look at four from the past several decades that will begin to reveal how and why bad unintended consequences have tended to arise in Big Tech.

We'll consider the personal computer, the internet, the smartphone, and social media. Some of this material may be familiar, but it will provide the means for us to start collecting problematic factors and assembling the spokes of our Wheel.

Chapter 1
"This Changed Everything": The Advent of Internet Advertising

"It was only a matter of time before someone made them pay for what they thought they were getting for free."
—Jennifer Egan, from her novel *The Candy House*[1]

THE PEOPLE WHO DEVELOPED THE PERSONAL COMPUTER BElieved it would make life better for everyone.

They were an odd, intriguing mix of hardware gearheads and software hackers, hippies and entrepreneurs. Some had gotten their start in universities and labs on the East Coast of the U.S., or in the Midwest, but by the mid-20th century, many had migrated to California.[a] The aerospace industry there had already attracted talented engineers to institutions such as the Jet Propulsion Laboratory in Pasadena, and young engineers could receive training at

---

[a] Again, I'll note that they were not all based in California, or in the U.S. For example, British computer scientist Tim Berners-Lee invented the World Wide Web while working at CERN, the European Organization for Nuclear Research, in Geneva, Switzerland.

S.T.E.M.-oriented universities such as Stanford and Cal Tech. In the 1960s, California also became the home base for a new human potential movement, spurred by proponents of the counterculture such as Stewart Brand, with his *Whole Earth Catalog*. At places like the Esalen Institute, people put a lot of energy into developing their psychological and spiritual capabilities, but the movement also appealed to engineers who believed that they could empower people through technology. Steve Jobs liked to talk about how computers could become "a bicycle for the mind."[2]

The development of the personal computer is often described as a feat achieved by a few geeky engineers working out of their garages, but that's simplistic. A great deal of the groundwork had actually been funded by national governments, as when, in the 1950s and 60s, the U.S. Department of Defense poured money into companies in California to support the development of the transistor and the integrated circuit.[b] (It even subsidized housing for tech workers.[3]) Funding and research also came from companies such as IBM and Xerox, which foresaw that cheaper, smaller, and more powerful computers would make all sorts of business tasks more efficient.

Still, the new technology provided a great outlet for engineering and coding whizzes who loved surmounting the technical challenges, as well as techno-hippies thrilled by the thought of building machines that would empower users to do all sorts of things, including writing their own software programs, publishing from their desktops, and forming new kinds of communities.

When I spoke with Robert Gardiner,[c] a former coder and

---

[b] The boom in electronic computation had actually begun during World War II, moved forward by work for defense industries by scientists such as Alan Turing and John von Neumann. (https://www.cs.cmu.edu/~tcortina/15292s17/lectures/WWIIAndTheAdventOfModernComputing.pdf)

[c] By request, this is a pseudonym, the only one I have used in this book.

company co-founder in the financial tech industry, he gave a great sense of the personal rewards:

> "When I was first exposed to computer programming, it was a revelation to me. You have that impulse to create something, and to see if you can come up with some insight, some deep nature of a problem that lets you implement a solution that may not be the obvious one, but that is elegant. That's what I was always looking for: the opportunity to think very deeply about a particular problem and to build something that both solved the problem, and revealed the deep structure of it. It's like woodworking—like, I built this beautiful table. The satisfaction of conceiving of that and building it is what got me into this. (And the fact that people would pay me to do it—that was pretty good too!)"[4]

Such early enthusiasm about computers carried over into excitement about the fledgling internet, the network devised to link the machines, which was first developed with another positive intention: to make it easier for academic researchers to share their work. Once the standards encoded in the World Wide Web furthered the ease of connection, the technology was adopted by millions of users, inaugurating an era of radical change.

Never before the advent of digital technologies had it been so easy, fast, and cost-free to copy pieces of information; never before the internet had it been so easy, fast, and cost-free to very widely disseminate them.

Steven Johnson, who was involved in early internet ventures such as the online magazine *FEED*, has said that, "The web had promised a new kind of egalitarian media, populated by small magazines, bloggers, and self-organizing encyclopedias; the information titans that dominated mass culture in the 20th century would give way to a more decentralized system, defined by collaborative networks, not hierarchies and broadcast channels."[5] During its earliest days, users began to form communities on cyber bulletin boards where they could share computing tips and chat about all sorts of other interests, from gaming to gardening to fanatical appreciation for the Grateful Dead.

"Idealism and utopian dreams pervaded the industry," recalls McNamee.[d] "The prevailing view was that the Internet and World Wide Web would make the world more democratic, more fair, and more free."[6]

As I mentioned, the U.S. military funded a lot of the initial research, largely through its Advanced Research Projects Agency. Explains journalist Ben Tarnoff:

> "While ARPA had wide latitude, it still had to choose its projects with an eye toward developing technologies that might someday be useful for winning wars. The engineers who built the internet understood that, and tailored it accordingly. That's why they designed the internet to run anywhere: because the US military is everywhere. [...] The reason the internet can work across any device, network, and medium—the reason a smartphone in Sao Paulo can stream a song from a server in Singapore—is because it needed to be as ubiquitous as the American security apparatus that financed its construction."[7]

Military needs, the idealism of hippies, and the problem-solving mindset of engineers were all important drivers, but we haven't yet touched on one of the most crucial factors that enabled the creation of the internet. Tech journalist Paulina Borsook, who witnessed its early rise, explained that:

---

[d] As Columbia University professor Tim Wu points out in his book *The Master Switch: The Rise and Fall of Information Empires*, this kind of idealism was also expressed by the early creators and users of radio: "In the magazines of the 1910s you can feel the excitement of reaching strangers by radio, the connection with thousands, and the sheer wonder at the technology." (p. 37)

"The aspect of the Arpanet that gets overlooked, [...] which is an indirect effect of it having been a government project, is that [it] did *not* have to make money, create return on investment, or demonstrate clear commercial utility. That it became clear within a few years that people just loved sharing information, quick as a bunny, through the wonder of electronics—turned out to be an insight that thousands of people would be able to make money off in decades afterwards. But the Arpanet was *sheltered* from commercial pressures; in fact, in its first fifteen years, commercial traffic was strictly forbidden on it. Venture capital would not have funded its revolution-for-the-hell-of-it. The guys who worked on it weren't necessarily thinking of its applicability to the Outside World, of markets, of quick payback. They had the time and space and freedom of a decade or more to argue and futz around and tinker and bicker for the joy of creating something they found fun and worthy, and ended up creating an institution of technology intersecting with people—the Net—that you can only admire."[8]

The Age of the Internet began with a period of great excitement—of *joy*, really—when it seemed that people could create an incredible new vehicle for decentralized, peer-to-peer sharing. Most amazingly, perhaps, it was—except for the cost of a computer, a little electricity, and a dial-up connection—totally free.

~

Sadly, this period of viewing the Web as a harmonious, noncommercial, and user-centered forum did not last long.

Legal scholar Tim Wu describes a trend that he calls the *Cycle*: "History shows a typical progression of information technologies: from somebody's hobby to somebody's industry; from jury-rigged contraptions to slick production marvel; from a freely accessible channel to one strictly controlled by a single corporation or cartel—from open to closed system."[9]

In the case of our new digital technologies, the results of this cycle have been stunning. As Steven Johnson put it in 2018, "Twenty years after the Web first crested into the popular imagination, it has produced in Google, Facebook and Amazon—and indirectly, Apple—what may well be the most powerful and valuable corporations in the history of capitalism."[10]

How was that possible? How could a medium based on free exchanges of information and ideas between users end up enabling a few entrepreneurs to make such vast fortunes?

For Amazon, the answer was to use the internet for selling products. For Apple, it was to use strong engineering and design—and clever marketing—to sell popular internet-connectable hardware.

But many internet-based companies—such as Google and Facebook—gave their services to users for *nothing*. How could they ever get rich by doing that?

As someone who has often conducted Google searches as many as fifty times a day, it pains me to write this next section. In the past, I thought of the company as a benevolent business. How could you not like a company that provides such an incredibly useful free service, and that had the motto *Don't Be Evil* written into its original corporate code of conduct?[e]

Originally named BackRub, believe it or not,[f] the company

---

was founded in 1998 by two Stanford Ph.D. students named Larry Page and Sergey Brin, who did a brilliant job of creating an internet search engine. That same year, they presented a conference paper in which they warned that, "We expect that advertising-funded research engines will be inherently biased towards the advertisers and away from the needs of the consumers,"[11] so they decided that their new website would be ad-free.

As users typed in queries, the company's workers gathered data about what they were looking for, and used that information to improve the search capabilities. Their product definitely worked. The problem, from the perspective of investors in the fledgling company, was that it wasn't making money.

Brin and Page had figured that their company might earn revenue by licensing their software to others. When they took on backing from venture capitalists, though, the hard-nosed businessmen didn't think that sounded like a way to bring in a big-enough return on investment. A common V.C. move in those days was to prod a startup to bring on "grownups"—mature, experienced businesspeople—to help its young visionaries run their companies. For Google, these were people such as Eric Schmidt, the company's CEO from 2001-2011. (In the case of Facebook, the chief adviser was Sheryl Sandberg.)

"Page and Brin had been reluctant to embrace advertising," notes Harvard professor Shoshana Zuboff, "but as the evidence mounted that ads could save the company from crisis, their attitudes shifted. Saving the company also meant saving themselves from being just another couple of very smart guys who couldn't figure out how to make real money, insignificant players in the intensely material and competitive culture of Silicon Valley. [...] Brin had his own take: 'Honestly, when we were still in the dot-

---

what other sites it related to." (https://www.businessinsider.com/the-true-story-behind-googles-first-name-backrub-2015-10)

com boom days, I felt like a schmuck. I had an internet startup—so did everybody else. It was unprofitable, like everybody else's'."[12]

For its first few years, the internet had not been seen as a way to make "real money"—or *any* financial profit. Only government agencies and universities could use it. From 1986 on, most of the traffic ran on a network, NSFNet, created by the National Science Foundation, where commercial uses were prohibited. The public was first allowed to access it at the end of the 80s, and its use was greatly accelerated by Tim Berners-Lee's invention of the World Wide Web in 1991. That was also the year when the NSF lifted its restrictions on commercial use,[13] and it didn't take long for entrepreneurs to start figuring out how to financially exploit the Web.

As the first internet service providers (such as CompuServe and MCI) and the first browsers (such as Mosaic and Netscape Navigator) began to gain users, sharp-eyed businesspeople saw two main opportunities: to use the Web as a vehicle for selling things,[g] and to use it as a vehicle for selling ads.

As we'll soon see, that second choice would have massive negative unintended consequences.

In 1993, a website called Global Network Navigator sold the first clickable display ad to a Silicon Valley law firm.[14] The next year saw the first commercial spam[h] message, when a couple who

---

[g] Journalist Jenny Kleeman has pointed to one group of entrepreneurs who found another way to use this new medium for making money, and made major contributions to its development: "Online pornography pushed the growth of the internet, transforming it from a military invention used by geeks and academics to a global phenomenon. Pornography was the motivator behind the development of streaming video, the innovation of online credit card transactions, and the drive for greater bandwidth." (From "The Race to Build the World's First Sex Robot," Apr. 27, 2017 in *TheGuardian.com*, https://www.theguardian.com/technology/2017/apr/27/race-to-build-world-first-sex-robot)

[h] This soon-ubiquitous use of the internet was thus labeled in homage to a hilarious sketch by the British comedy troupe Monty Python.

practiced immigration law in Arizona emailed an ad to 6,000 usegroups.[15] The new advertising market spurred a commercial frenzy that has often been likened to the California Gold Rush. In August of 1995, before it had even earned a dime of profit, Netscape made an initial public offering on Wall Street, and it was valued at 2.9 billion dollars. The Dot-com Boom was on, a period of financial speculation that raised the value of internet-based companies to wild (and often absurd) heights. Early Google employee Heather Cairns recalled what those days were like: "*Going public. Being rich. Going public. Going public.* That was really on the forefront of so many people's minds."[16]

"After a while, the whole thing became more sharp-elbowed," remembers former Facebook product manager Antonio Garcia Martinez. "It wasn't hippies showing up anymore. There was a lot more of the libertarian, screw-the-government ethos. That whole idea of move fast, break things, and damn the consequences."[17]

Prior to this moment, advertising had been around for quite a while, of course, but its use had been relatively scattershot. An ad on network TV sprayed out to millions of people—to potential customers, it was hoped, but also to many viewers who might have no interest in the product. It was possible to focus the reception of the ads more than that—a company that sold guitars, for example, could place an ad in a guitar players' magazine—but many companies had to drop their lures in front of big, undifferentiated masses of people, and hope for the best.

Beginning in 2000, business-minded executives at Google realized that, while conducting their searches, their users were unwittingly providing reams of data that went far beyond the information needed to improve the search process. They were revealing where they lived, what their hobbies were, their preferences and affiliations. The execs realized that this "surplus" data could be exploited to give the company a whole new profit center: it could promise advertisers that each user could be shown enticements specifically tailored to their tastes and interests. To people in the ad business, this seemed like the fulfillment of their wildest dreams, and it soon became the core of Google's business, enabling a powerful new economic engine that Professor Zuboff has termed *surveillance capitalism.*

By now, the technology has grown remarkably sophisticated, using a technique called *programmatic advertising,* which was created by Brian O'Kelley, who was the chief technical officer of the Right Media Exchange, a platform for targeting audiences and monetizing digital media. When you click on a web page, he explains, that action generates a *cookie,* a little text file that identifies your specific computer. Then an auction is held. The platform selling the ad space sends the cookie "out to maybe a hundred, maybe more, different firms, and they say, 'How much will you pay to buy this ad spot for this cookie right now?' They are then merging that with any historical data they know about you that they've purchased from various vendors, [...] doing anything they can to match it with something that tells them more about you. [...] All this is happening, in this blink of an eye, between when the website loads and the ad loads."[18,i] (In other words, when you and I load the same web page on our different devices, we may be presented with completely different ads. If you have kids, they may see different ads as well.)

This information vacuumed out of users was enormously valuable to Google, and to other companies that were able to extract it.

---

i   When we're looking at who is making such surveillance possible, it's easy to just point fingers at corporations such as Facebook and Google, but Antonio Garcia Martinez argues that "the role of modern-day Big Brother is actually played by companies you've likely never heard about. Companies with names like Axciom, Experian, Merkle, and Neustar (among others)." (*Chaos Monkeys.* P. 384)

For example, tech journalist Geoffrey notes that "in 2013, the average American's data was worth about $19 per year in advertising sales to Facebook, according to its financial statements. In 2020, your data was worth $164 per year."[19] By 2021, the company saw its market valuation rise above a trillion dollars.[20]

As they developed this financial model, the Big Tech companies made a major conceptual shift. Originally, there had been an implied contract between the companies and their users that said, in effect, *You share your data with us and we'll use it to improve your experience on our site.* That's what good companies do: they serve their customers. But Google or Facebook users are not really "customers," since they don't pay anything. Companies that buy ads *do*.

"This new market form," says Zuboff, "declares that serving the genuine needs of people is less lucrative, and therefore less important, than selling predictions of their behavior.[j] [...] This changed everything."[21]

Let's start building our Wheel. The use of advertising as the main source of revenue for a digital platform is one factor.

The demotion of users from the center of a company's considerations to the periphery is another.

Advertisers pay tech companies when users click on their ads. This means of earning is known as *PPC* marketing, for "pay-per-click," and it has hugely shaped the content that now floods the Web. Documentary filmmaker and activist Astra Taylor points out that, "The dominant economic model compels the quest for hits and clicks, encouraging content farm logic—search engine-optimized headlines, page-view-baiting slide shows, irresistibly shared

---

[j]   That is, the tech companies promise advertisers that the personalized targeting would make users more likely to buy.

top-ten lists, image-heavy posts, bite-size webisodes, TV-style video promotions, and so on."[22]

As Jeff Hammerbacher, a software coder and one of Facebook's earliest hires, succinctly put it, 'The best minds of my generation are thinking about how to make people click ads. That sucks.'"[23]

The promotion of trivial content in order to lure people into clicking on ads may seem like a trivial problem, but as we're about to see, the choice to make advertising the internet's engine of profit had a number of dire consequences.

First, though, let's look at another reason why the ad-based model has shaped the modern internet. Personalized targeting was one way to raise the PPC metric, but another was to boost the overall number of viewers for the ads. Social media platforms provided an amazingly cheap and effective way to do that.

By way of contrast, let's look at an older business model. Sears, Roebuck and Co. was founded in 1893, and at first it operated as a mail-order catalog enterprise. It took 18 years for it to become a public company, 32 years before it began opening retail stores, and about a century before it became the largest retailer in the U.S.[24] Along the way, it had to rely on the old-fashioned infrastructure of physical mail delivery, and then it had to expand physically, store by store, across the nation. For Sears to grow into a massively successful company, it needed to hire an enormous number of employees, from mail order fulfillment teams to store clerks to corporate office workers.

Facebook, founded in 2004, gained six hundred million users within six years,[25] and almost three billion by 2023.[26] And because social media companies relied on the internet, rather than physical factories or stores, once they put their basic software in place, they could generate wild increases in the number of users with little rise in costs, including the cost of labor. (WhatsApp, for example, had a total workforce of just 55 people when Facebook acquired it in 2014—for 19 billion dollars.[27])

These companies were relatively easy to grow because they were based on building networks of users, and they benefited from a principle known as Metcalfe's Law, which says that a network's value increases exponentially with the number of people in it. That influenced the business model of the budding Silicon Valley, in which **investors put a lot of pressure on companies to show extreme growth—a major spoke within our Wheel**.

Big Tech founders found ingenious ways to get their users to do much of the work of creating that growth. Remember Tom Sawyer and the famous episode in which he convinced his pals to help him paint a fence for free? Mark Twain would have been astounded to see how social media companies have advanced the technique.

Every time we ask someone if they'll be our Facebook friend, or we follow someone on another platform, we're helping a fabulously wealthy company build its user network, and we're doing that labor for free.

Every time we post something on Instagram or TikTok, we're creating content for the platform, and we're doing that labor for free.

Every time we share someone else's post on any of these platforms, we're helping the company distribute its content, and we're doing that labor . . . you guessed it . . . for free.

Chapter 2
Clicks And Consequences: Psychological And Social Effects

Welcome to the internet
> *What would you prefer?*
> *Would you like to fight for civil rights or tweet a racial slur?*
> *Be happy*
> *Be horny*
> *Be bursting with rage*
> *We got a million different ways to engage*
> —Bo Burnham, from his song "Welcome to the Internet"[1]

As if extracting our data and exploiting our labor weren't enough, the Big Tech companies have done one more big thing to suck value out of us.

In my gym, I watch many patrons compulsively check their smartphones in the seconds between every workout set. I have gone to a Broadway show and seen the audience light up like a field of fireflies the second the intermission started, and then watched the ushers have to go around and literally beg people to turn their phones off—*after* the next act began. I stood on the edge of a lake in the middle of Brooklyn's gorgeous Prospect Park and saw a bunch of rented paddle boats drifting aimlessly because all the people aboard were too enraptured by their phones to paddle or steer.

Aside from providing advertisers with vast networks of individually targeted users, social media companies can also promise that they'll keep those users riveted to the websites and apps displaying their ads. This is a book about unintended consequences of technology, but perhaps no consequence has ever been more fervently intended. Sean Parker, the first president of Facebook, said this quiet part out loud when he admitted that:

"The thought process that went into building these applications, Facebook being the first of them to really understand it, that thought process was all about: 'How do we consume as much of your time and conscious attention as possible?' And that means that we need to sort of give you a little dopamine hit every once in a while, because someone liked or commented on a photo or a post or whatever. And that's going to get you to contribute more content, and that's going to get you more likes and comments. It's a social-validation feedback loop, exactly the kind of thing that a hacker like myself would come up with, because you're exploiting a vulnerability in human psychology. The inventors, creators—it's me, it's Mark, it's Kevin Systrom on Instagram, it's all of these people—understood this consciously. And we did it anyway."[2,a]

---

[a] Over the past couple of decades, thousands of Stanford students have taken a course given by social scientist B.J. Fogg called *Persuasive Design*, in which he reveals how to use psychological techniques to get people to do what you want—and then those young people have gone on to careers all over Silicon Valley, where they embedded those techniques into our devices. Another prominent teacher and

To boost our attention (and thus their ad sales), they added a host of features designed to keep us glued to our screens, such as the red dots that tell us we have a new message (2001), the Like button (2009), and push notifications from apps (2009). And who among us can resist clicking on a link after we've been told that we've been photo-tagged (2011)? These days, not only are we falling into our phones, we're never hitting bottom. That boundless appetite was encouraged by Aza Raskin, the interface designer who invented infinite scrolling (2006).[b]

The Share button (2010) gave companies a unique economic twofer. As journalist Chris Hayes puts it, "I view and then I share. I view and then I share. [...] The genius of this innovation from a business standpoint is that it creates two opportunities for attention to be captured and monetized."[3]

The lure of all these features is so powerful that our phones seduce our attention even when they're not on. In one study, researchers observed one hundred pairs of test subjects engaged in conversation. Half had their phones visible nearby. Even though they didn't use them during the experiment, the authors found that, "The mere presence of mobile phones inhibited the development of interpersonal closeness and trust, and reduced the extent to which individuals felt empathy and understanding from their partners."[4]

My students glance anxiously at their phones when they place them in view, and that's why I ask them to stow them away. It's not hard to see why they have a hard time complying: they've been subjected to all sorts of psychological tricks devised to keep them fixated on their screens. And they *are* fixated, even if that doesn't feel enjoyable or satisfying.

---

tech consultant is Nir Eyal, who actually wrote a book called *Hooked: How to Build Habit-Forming Products*.
[b] Years later, repentant, Raskin became an activist for the ethical design of tech.

The designers of the attention-grabbing features achieved their intended result, but it has had all sorts of unintended effects.

For one thing, our new digital technologies fragment our attention. We may believe that we're able to focus on more than one thing on a time, such as texting while working on a project, but researchers have shown that such bouncing around renders us less efficient and more likely to make mistakes.[5]

On the other hand, there are times when lack of focus can be a good thing. As a writer, some of my most productive moments come when I'm doing some simple physical activity, like taking a shower or strolling aimlessly, and I can let my mind roam free. I couldn't write novels if I didn't daydream. As a writing teacher, I worry about my students who are staring at their phones in every free moment: are they giving their subconscious minds the chance to percolate ideas?

Natalie Goldberg, a Zen Buddhist best known for her book *Writing Down the Bones*, wrote about attending a luncheon at which she chatted with a bunch of young software engineers:

> "In truth, I was disappointed that all of this technology was discovered in my lifetime. It seemed to make time busier, more complicated, as if the functions of the mind, the beat of thought I'd come to depend on for my years of sitting and writing practice, no longer applied. I understood how the brain made poetic leaps, how it could juxtapose seemingly dissimilar objects, people, rivers, fruit, how you could reach into the center of the source and discover a vast emptiness that was full and abundant. But the rhythm of the minds partaking of this meal felt jagged, even severed in some places, as though natural mind waves had been broken."[6]

Another manifestation of change in our habits of mind is the fact that—as anyone who has glanced at Instagram or TikTok recently can attest—our digital technologies have enabled a remarkable blooming of narcissistic behavior, as with the invention of the user-facing phone camera.[c] Soon after, developers began creating digital "beauty" filters, and now social media is inundated with the phenomenon of "influencers," egomaniacs who bombard users with images of themselves.[d] That's part of the reason why social media has proved especially damaging for adolescent girls: the flip side of self-admiration is the anxiety that comes from constantly comparing your looks with those of others—and constantly feeling that you can't measure up.

Boys are vulnerable too. Social psychologist Jonathan Haidt has noted that though the smartphone was originally designed as "a digital Swiss Army knife that you pull out when *you* need something," with the advent of social media, it became "a portal that [...] that companies now can use to get to you, as a child. Without your parents' permission or knowledge, they can get to you. They can send you notifications. They can try to get you to stop your homework and: *Come—look at what someone just said about you.*"[7]

Indeed, social media has proved a perfect medium for bullying. Computer scientist Jaron Lanier says that when he talks to young people about it, he is "always struck by the endless stress they put themselves through. They must manage their online reputations constantly, avoiding the ever-roaming evil eye of the hive mind, which can turn on an individual at any moment. A 'Facebook generation' young person who suddenly becomes humiliated online has no way out, for there is only one hive."[8]

---

[c]   The first one was released in 1999, although the selfie trend didn't really take off until the 2010 introduction of the first Apple phone with one, the iPhone 4.

[d]   I can only find one positive thing to say about these people, which is that at least they have gotten social media platforms to share some of the profit derived from user-posted content.

In his book *The Anxious Generation: How the Great Rewiring of Childhood Is Causing an Epidemic of Mental Illness,*[9] Haidt lays out a number of other problems with smartphones and social media, including how they make pornography easily available to children who are too young to deal with it, and how they encourage them to spend so much time alone, where they're unable to develop valuable social skills that have traditionally been gained through direct play with other kids.

Haidt presents evidence showing a particularly large spike in adolescent mental illness between 2010 and 2015. What else happened during that period? It saw major increases in the availability of high-speed internet,[10] the percentage of children owning smartphones,[11,e] and kids' use of social media. Silicon Valley leaders had introduced a slew of technologies that cycloned together in a perfect storm that's still battering the psyches of children around the world today.

The developers of Big Tech are also affecting our brains in other ways, because the competition for our attention is so intense that they're finding increasingly extreme ways to win it. If you use social media, you may have noticed that your feeds show you videos of people suffering extreme falls, or making offensive statements, or fighting with each other in airplane aisles. I look at those videos and find myself wincing or gasping, or feel my blood begin to boil.

That's no accident. Roger McNamee explains that, "Getting a user outraged, anxious, or afraid is a powerful way to increase

---

e With desktop computers, kids only spent a limited amount of time on them, at home; with portable smartphones, they can access the internet anywhere and anytime. (As many teachers have been chagrined to witness.)

engagement. Anxious and fearful users check the site more frequently. Outraged users share more content to let other people know what they should also be outraged about."[12]

Others are nudged toward harming themselves. In 2021, a Facebook internal investigation showed that Instagram's algorithms pushed girls looking for fashion and diet tips toward content promoting eating disorders such as bulimia and anorexia.[13,f,g] This didn't happen because anyone who worked at Facebook had any intention of harming adolescent girls; it was simply a function of automated computer algorithms embedded in the app. One said, *If a user looks at posts about* this, *show them more posts about it.* A more dangerous one directed: *If a user looks at posts about* this, *show them more extreme posts about it.*

And this is another spoke in our Wheel: algorithms that automatically push users toward more intense engagement with a platform or product can push us toward bad unintended consequences.

A 2023 lawsuit provided evidence that Mark Zuckerberg ignored urging from some of his own top executives to do more to address their role in problems such as bullying, harassment, and teen suicide.[14] **This is also a common Wheel trait: a willingness of tech leaders to ignore evidence about how their products are causing (or might cause) harm.**

These apps assail adult psyches too. The amazing thing is that we stare at our screens even though our time on them often leaves us feeling us bad. Psychologist Katy Cook notes that "study after study tells us that social media is negatively correlated with well-being and life satisfaction, and is associated with increased levels

---

f   Facebook (now Meta) owns Instagram.

g   Of course this is just anecdotal evidence, but one of my college students estimated that 70 percent of the girls in her high school had suffered from eating disorders, and she attributed this to the negative effects of TikTok and other social media platforms.

of negative emotions, loneliness, and depression."[15] And though the time we spend on the Web can divert us from some worries, all too often it "leaves us constantly on alert, suffering from heightened cortisol levels, and feeling increasingly overstimulated, anxious, and distracted."[16]

Receiving a Facebook or Instagram Like *does* make us feel good, at least for a few seconds, because it rewards a dopamine-inspired desire,[h] but Cook makes an important distinction between this hormone and serotonin, which is stimulated by things like exercise, a healthy diet, sunlight, and human touch: "Dopamine is addictive, short-term, visceral (meaning it's felt in the body), self-perpetuating, and is generally a solitary experience. Perhaps the most important feature of dopamine is that it always makes us want more dopamine. Serotonin, by contrast, is non-addictive, long-term, ethereal (meaning it's experienced in the mind, not the body), is generally a shared experience, inspires leadership and generosity, and is associated with contentment."[17]

When Big Tech companies use extreme techniques to monopolize our attention and addict us to their products, they add another spoke to our Wheel. They keep us glued to their platforms so they can sell our attention to companies that want to show us ads. While we're there, they extract data from us so that they can get better at targeting more ads. It's a vicious cycle: *Draw our attention in; suck our data out.*

Despite the evidence of harmful effects, many of us still feel that we can ignore these operations. Every time a newspaper runs

---

[h] Contrary to a popular misconception, dopamine doesn't deliver the actual burst of pleasure, but rather boosts a craving for such rewards, and the motivation to seek them out. (https://mhanational.org/what-dopamine)

an article about how Big Tech companies are forcing our attention and invading our privacy, I scan the reader comments and find that many users don't see cause for alarm. *Why should I care if they're looking at my data?* one will write. *I'm not doing anything I need to hide.* Or, *We're grownups here. If I don't like Facebook, all I have to do is quit it.*

If only it were that easy. Tech journalist Geoffrey Fowler spent some time looking into how Facebook collects data. Pretty much everyone now knows that it reaps information from you when you're on its site. But the company "is so big," he wrote in 2021, that:

> "It has persuaded millions of other businesses, apps, and websites to also snoop on its behalf. Even when you're not actively using Facebook. Even when you're not online. Even, perhaps, if you've never had a Facebook account. Here's how it works: Facebook provides its business partners tracking software they embed in apps, websites, and loyalty programs. [...] Among the 100 most popular smartphone apps, you can find Facebook software in 61 of them. [...]. Facebook also has trackers in about 25 percent of websites. [...] Over two weeks, Facebook tracked me on at least 95 different apps, websites and businesses, and those are just the ones I know about. It was as if Facebook had hired a private eye to prepare a dossier about my life."[18]

Many digital features that are presented to us as conveniences are actually Trojan Horses: hidden inside are means for companies to track us. For example, when we can use a Facebook or Google button to sign in to some unrelated website, we think, "How nice that they're providing this helpful feature!" In reality, they're expanding their data-grabbing reach.

Let's mark this as another characteristic of the Wheel: when tech executives promise that their latest

technology will offer users greater convenience, they're often planning to use that technology to extract something from us, such as personal data or unpaid labor.

It's not just Facebook and Google who are tracking us, and it's not just happening on the Web. Big Tech's data-gathering operations are oozing beyond the confines of the internet, into what has traditionally been known as the "real world." Many tech entrepreneurs are turning their attention to a whole new source of data gold: what's known as the *Internet of Things*, which connects all sorts of physical objects to digital monitors. They present this with the promise of greater efficiency, but even objects that are supposed to make it easier to sleep better (SleepNumber beds) or control the temperatures of our homes (Nest thermostats) or provide us with entertainment (smart TVs) can enable the surreptitious gathering of information about our behavior. In 2017 the iRobot company made news when its CEO told *Reuters* that it planned to make extra revenue by having its Roomba vacuum cleaners map floor plans of customer's homes, which it could sell to other companies.[19]

When the companies that make these devices sell our personal data to advertisers, we're often oblivious that this is happening; if we're told, the information is often buried deep inside legal privacy notices that are too long and dense for most people to deal with. (When was the last time you read one of those all the way through?) We're also unaware because the transactions are happening in such a frictionless way. As computer scientist Dipayan Ghosh puts it, "Customers do not experience the collection of their data or the sale of their attention; they have simply downloaded all of their favorite apps, apparently for free."[20]

To potentially make matters worse, A.I. chatbots have already become incredibly skilled at gathering and synthesizing information from data sets such as the contents of the internet. When we interact with them, they are applying the same techniques to us. As they get to know us, they become increasingly chummy, overcoming our defenses as though we're confiding in a friend. As journalist Vauhini Vara argues, "The stage is set, then, for the next phase of Big Tech's ever-deepening exploitation of the natural human desire for information, connection and well-being."[21] (Not to mention the ultimate micro-targeting of ads.)

As if all of this were not alarming enough, tech engineers are now developing a technology called *affective* computing, which will enable our phones and other devices to scan our faces and/or bodies in order to discern our emotional states and identify when we might be particularly vulnerable to sales pitches.[22]

Until recently, I smugly thought, *I don't care if Facebook or Google shows me ads. I'm the kind of person who doesn't pay attention to advertising.* And then I decided to buy some apartment renter's insurance, and noticed that the only companies I was considering buying it from were the ones that had been barraging me with TV commercials for the past few years. Humbled, I looked at my underwear drawer (hello Calvin Klein), my medicine cabinet (hello Colgate) and my refrigerator (hello Heinz ketchup). I'm sure there are small companies that make even better toothpaste and ketchup, but I didn't buy from them because they couldn't compete with the ad budgets of the mega-corporations.

If I have proved so susceptible to relatively crude TV ads, how well am I going to be able to resist when companies use facial recognition technology to see when I look sad, and take the opportunity to sell me something sweet?

These attempts to influence us are only getting more extreme. Not content to simply collect data, Big Tech executives are now focusing on "actuation," which is designed to push us toward certain *behaviors*. Have you noticed that when you visit a new website these days, it will often ask if it can have permission to access your location—even if that has nothing to do with the functioning of the site? That's because location data can be sold to other companies, and it has become increasingly valuable. If those companies know where we are, they can send us ads as we pass stores to urge us to enter, and soon they may even be able to activate personalized holographic ads that will seem to materialize around us.[23]

"Your smartphone can broadcast your exact location thousands of times per day, through hundreds of apps, instantaneously to dozens of different companies," noted a *New York Times* op-ed titled "Total Surveillance Is Not What America Signed Up For." "Each of those companies has the power to follow individual mobile phones wherever they go, in near-real time. That's not a glitch in the system. It is the system. If the government ordered Americans to continuously provide such precise, real-time information about themselves, there would be a revolt."[24]

Imagine what a future Stalin or Mao could do with that level of monitoring.

Actually, we don't have to imagine bad political effects of these technologies.

They're already here.

Chapter 3
Cozy Communities And Creepy Clusters: Political Effects

"Perhaps the greatest social effect of radio-communication, so far, has been a political one: the restoration of direct contact between the leader and the group. Plato defined the limits of the size of a city as the number of people who could hear the voice of a single orator: today those limits do not define a city but a civilization. [...] The possibilities for good and evil here are immense. [...] *As with all instruments of multiplication the critical question is as to the function and quality of the object one is multiplying.* There is no satisfactory answer to this on the basis of technics alone: certainly nothing to indicate, as the earlier exponents of instantaneous communication seem pretty uniformly to have thought, that the results will automatically be favorable to the community."

—Lewis Mumford, in 1934[1]

As I mentioned, a number of early developers of the internet thought that they were creating a boon for humanity, a way to spread knowledge, reason, and democracy around the world. They believed in the notion that sunlight is the best disinfectant. (How, for example, could a dictator maintain a regime founded on lies if his citizens could easily access the truth online?) As they built an amazing network of Roman roads to ease the passage of good information, what they failed to foresee was that they were also enabling the vast reach of bad info. A huge unexpected consequence of their great invention has been—as we can all plainly see—a world that seems to be spiraling into profound *un*reason.

If, for example, you could have told me several decades ago that in 2021 almost a quarter of U.S. Republicans would end up believing that an enormously powerful cabal of cannibalistic Satanists, including Tom Hanks and Oprah Winfrey, were operating a global child sex trafficking ring,[2] I would have felt certain you were joking.

Today, of course, no one is laughing.

Here's another example. 675,000 Americans died in the great influenza pandemic of 1918.[3] Before Covid-19 hit the U.S. in 2020, I felt sure that, given a century's worth of stunning advances in medical science, we'd fare much better against a new contagion —yet in the first five years of the pandemic almost twice as many U.S. citizens died from the new virus.[4] The toll would have been significantly lower[5] if so many people had not been encouraged— largely through misinformation spread on social media—to avoid sensible health measures because of wild conspiracy theories, such as the notion that Bill Gates was using the Covid vaccine to implant trackable microchips into the population. (A May 2020 poll of U.S. adults showed that a whopping 28% believed this ridiculous theory.[6])

Of course, eccentrics have floated crazy ideas throughout human history. In the last century, a few believed that the Apollo moon landings were faked by film director Stanley Kubrick,[7] or that the British royal family were actually shape-shifting alien reptiles,[8] but such theories were limited to small fringe groups. In the past few years, though, preposterous notions have been taken up by tens of millions of people worldwide, and they've gone from being relatively harmless to extremely impactful, affecting how health crises are dealt with (or not), who citizens vote into power, what policies get enacted, and what policies don't.

How is this possible? How can it be that in our very high-tech era, our supposed age of advanced science and rationality, vast numbers of citizens have taken up beliefs that are so obviously and demonstrably nuts?

Every semester, I teach a course called Critical Thinking and Writing. I'm glad to do so, because I hope to help students inoculate themselves against this epidemic of *un*critical thought.

I ask if they believe that humans think and behave rationally—and, if not, what might cause us to do otherwise. Then we examine how humans are subject to a whole range of biases that can distort our thinking.[a]

For example, we may be subject to *anchoring bias*.[9] If I ask how many people get killed annually by sharks, and we go around the table and nine students give an answer in the thousands, the tenth will likely also answer in that (inaccurate) range.

Clearly, we can be heavily influenced by peer pressure. Another example is *groupthink*, "the psychological phenomenon that occurs within a group of people in which the desire for harmony or conformity in the group results in an irrational or

---

[a]   The classic text on such cognitive biases is *Thinking, Fast and Slow* by Nobel-prize-winning psychologist Daniel Kahneman, who researched them with his colleague Amos Tversky. Over and over, they found ways in which humans *don't* think or behave rationally.

dysfunctional decision-making outcome. Group members try to minimize conflict and reach a consensus decision without critical evaluation of alternative viewpoints by actively suppressing dissenting viewpoints, and by isolating themselves from outside influences."[10]

When we explore conspiracy theories, such as the belief that the Earth is flat, we see how even more biases come into play. Among these are *confirmation bias* ("The tendency to search for, interpret, focus on, and remember information in a way that confirms one's preconceptions") and the *illusory truth effect* ("the tendency to believe that a statement is true if it has been stated multiple times").[11,b]

Do these penchants for distorted thinking sound familiar to you?

In the past few years, I have seen one friend get snared in a maelstrom of conspiracy theories (including 9/11 "trutherism"), and another who got enraptured by extreme political ideology. You likely know at least one person who has done the same. We might say that these people "went down a rabbit hole," but what does that really mean? And why does it happen?

Literally, the phrase comes from British mathematician Lewis Carroll's 1865 book *Alice's Adventures in Wonderland*, in which the title character follows a rabbit down his hole, at which point she suddenly passes from a world governed by logic into a highly irrational one. In real-life cases, though, the transition is usually not instantaneous; it's a gradual process—and it's often facilitated by our digital technologies.

Let's examine how this works.

---

[b] Donald Trump, who learned this technique from his mentor Roy Cohn, has repeated his lie about the 2020 election being stolen hundreds—maybe even thousands—of times.

<u>The Ability to Connect in Small Private Groups</u>

Homo Sapiens have always sought to create communities, whether in clans, tribes, or nations. Social media relies on that deep impulse, and that's not necessarily a bad thing. In fact, I have a confession to make: despite the many times I criticize Mark Zuckerberg in this book, I'm still on his platform. To give credit where it's due, his statement that he wants to "give people the power to build community" *is* borne out in some ways on Facebook. I use the site because it provides an easy way of seeing what old, distant friends are up to, and I'm a member of a Group that allows me and my neighbors to share local news and restaurant tips, volunteer for events, and even find lost pets. The inventors of the internet would have been proud of this fulfillment of their benevolent goals.[c]

Unfortunately, the company also has less altruistic motives in facilitating communities. Stanford Professor Adrian Daub notes that "Facebook loves Groups because they enable easy targeting by advertisers," and he adds that, "Bad actors like them for the same reason."[12]

Author Cory Doctorow points out that the feature makes it easy for people such as neo-Nazis and white supremacists to connect with like-minded people; once they express interest in extreme content, the platform's algorithms steer them toward such groups. And while social disapproval would normally make it difficult for them to advertise in public, they can do so covertly via Group-targeted ads.[13]

In 2016, a Facebook researcher named Monica Lee gave an internal presentation in which she warned that the platform was

---

[c]   I'll add that if I steer clear of the creepy political stuff on YouTube, I have found the educational opportunities to be pretty great, from learning how to play Delta blues guitar, to how to unclog a sink, to the best ways to cook salmon.

hosting a large number of such creepy clusters—and she found that "64% of all extremist group joins are due to our recommendation tools."[14]

In 2021, Scot Seddon, the national commander of the far-right militia group American Patriots Three Percent, put this enabling in even starker terms when he told his followers, "Facebook has been our greatest weapon. It's gotten us where we are today."[15,d]

<u>The Ability to Connect Through Giant Public Networks</u>

Social media platforms can provide community for lonely individuals who feel disenfranchised from society, including budding jihadists and spiteful incels.[e] Online, they can easily connect with small underground groups.

Digital networks also enable users to connect in huge public groups. Before the internet, someone who wanted to spread a false notion would have been hard-pressed to share it beyond their immediate physical circle, but one poster can disseminate a piece of bogus information to millions of people around the world with just one click, and the Big Tech companies will commonly do nothing to question or block it.

Wheel factor: tech that enables effortless, instantaneous, unmonitored, and unvetted communication through vast networks can enable an unprecedented dissemination of false, often dangerous ideas.

People who have built their own networks within the platforms are in an especially strong position to broadcast dangerous

---

[d] In 2024, Meta made a push to remove pages and accounts that were focused on recruiting for AP3 and other militia organizations. (https://www.propublica.org/article/inside-secret-ap3-militia-american-patriots-three-percent)

[e] "Involuntary celibates"

information, as when a number of TikTok influencers told millions of their followers in 2024—at that point, the hottest year in recorded history[16]—that they didn't need to wear sunscreen or sunglasses.[17]

A small number of people can have a hugely disproportionate impact on the spread of bad information. Researchers of the 2016 U.S. election found that just 0.1% of users were responsible for sharing approximately 80% of the political misinformation on Twitter.[18] Another study, in 2021, found that just twelve users, which it called "the Disinformation Dozen," were responsible for 65% of anti-vaccine content being spread across social media.[19] (Hello, Robert F. Kennedy Jr.)

In sum, the network effects of social media have made it easy for great masses of citizens to plunge down rabbit holes together.

<u>The Addition of Special Features</u>

I imagine that it would be difficult to sustain a belief in flat-Earthism or cannibalistic Satanists on one's own, but in the past couple of decades, Big Tech developers provided powerful rein-forcement for people who hold such extreme beliefs. One of our deepest needs is for the approval of our social groups, and they added several features that enable us to get it.

I rarely post on social media, but when I do, I'm shocked by how I crave Likes and other positive responses—and how good I feel when they appear. If I can get a warm feeling when my neigh-bors Like my restaurant tip, it's not hard to imagine how, if I were a neo-Nazi, I might enjoy that same fuzzy feeling by collecting Likes for some vicious anti-immigrant rant.

The Share and Retweet buttons provided a turbo boost for anyone who wanted to spread false information and collect those Likes. True facts couldn't keep up: during the 2020 U.S. election, researchers found that misinformation got six times the Likes and Shares on Facebook as news from reputable sources.[20]

<u>Extreme Algorithms</u>

As I mentioned earlier, the developers of social media platforms also added features that push users in more radical directions. Sociologist Zeynep Tufekci was doing research on YouTube when she noticed that if she watched videos of Donald Trump rallies, the platform's automated algorithms began to recommend and auto-play videos that featured "white supremacist rants, Holocaust denials, and other disturbing content."[21,f]

I saw this extremist effect for myself just after the 2024 presidential election, when a conservative 22-year-old I know sent me links to a couple of politically oriented YouTube videos. I had assumed that the push to extremes would be a gradual process, but after I watched just those two videos, my recommendations feed was immediately inundated with inflammatory ultra-right content.

It would be a big mistake to underestimate the effect of these algorithms. "Given its billion or so users," Tufekci has concluded, "YouTube may be one of the most powerful radicalizing instruments of the 21st century."[22]

Please note that these automated systems are utterly amoral: they make no distinction between promoting a cat video and an ISIS beheading[g]—and sometimes, evidently, tech executives don't either. "It's really sad," says digital media entrepreneur Rich

---

[f]   This push toward more extreme content also applied to nonpolitical topics: "videos about vegetarianism led to videos about veganism; videos about jogging led to videos about running ultramarathons."

[g] "In 2015, ISIS supporters had over 46 thousand accounts on Twitter, and posted 90,000 tweets a day. In 2013 there were more than 35 thousand ISIS videos on YouTube." (Jonathan Taplin, *Move Fast and Break Things: How Facebook, Google, and Amazon Cornered Culture and Undermined Democracy* (Back Bay Books: 2017) p. 183)

Kyanka, "because you have hundreds and hundreds of bigwig, smarty-pants guys at all these analytic firms, and they're trying to drill down into the numbers and figure out what kind of goods and products people want. But they only care about the metrics. They say, 'Well, this person is really interested in AR-15s. He buys ammunition in bulk. He really likes Alex Jones.[h] He likes surveying hotels for the best vantage points.' They look at that and they say, 'Let's serve him these ads. This is how we're going to make our money.' And they don't care beyond that."[23]

Filter and Preference Bubbles

Many observers have noted that the U.S. political scene is extremely polarized, perhaps more than at any time since the Civil War.[24] Of course, the recent extreme divisions in the nation—and other world democracies—didn't begin with the internet or social media; for decades before that, we've had partisan talk radio and cable TV. But a radio DJ or TV producer has to make a conscious choice to air disinformation. With social media, that's no longer the case: if users have looked at mis- or disinformation, algorithms will automatically show them more of it—and less solid information that might counter it. The result is what's known as a *filter bubble.*

Roger McNamee does a great job of combining all of these factors to explain why social media so powerfully enables polarization and irrational beliefs. In its digital silos, he argues:

"Users create an alternative reality, built around values shared with a tribe, which can focus on politics, religion, or something

---

[h] The radio show host who promoted the horrific conspiracy theory that the mass shooting of children at the Sandy Hook Elementary School in Newtown, Connecticut in 2012 had been faked.

else. They stop interacting with people with whom they disagree, reinforcing the power of the bubble. They go to war against any threat to their bubble, which for some users means going to war against democracy and legal norms. They disregard expertise in favor of voices from their tribe. They refuse to accept uncomfortable facts, even ones that are incontrovertible. This is how a large minority of Americans abandoned newspapers in favor of talk radio and websites that peddle conspiracy theories.[i] Filter bubbles and preference bubbles[j] undermine democracy by eliminating the last vestiges of common ground among a huge percentage of Americans. The tribe is all that matters, and anything that advances the tribe is legitimate."[25]

It would be easy to dismiss conspiracy theorists as intellectually deficient. Yet as public health researcher Devi Sridhar points out, "Those who believe outlandish theories are generally people who think of themselves as more intelligent than the average person, have a lot of time to do their own research on the internet, and are convinced that everyone else is being duped."[26]

Chris Hayes points out an obvious but surprisingly under-discussed aspect of conspiracy theories, which "is that they are often, in fact almost always, more attention-grabbing than reality. In evolutionary terms, shocking bits of false information outcom-

---

[i] Actually, many Americans *didn't* abandon newspapers: the rise of the internet put their local papers out of business. For at least a century, print media was supported by print advertising, including a section at the back of many publications for classified ads. The Web-based company Craigslist, founded by Craig Newmark, began to hoover up a great deal of that revenue after his site went national in 2000, and then other digital companies such as Facebook and Google diverted more money from newspaper coffers. Between just 2006 and 2011, their share of advertising revenues fell by more than half. (Taylor, p. 79)
[j] Filter bubbles are imposed, whereas preference bubbles are actively chosen.

pete mundane bits of true information."[27] It can be a lot more exciting—and perversely *fun*—to believe that the result of the 2020 U.S. presidential election was determined by a sinister, all-powerful global Satanist cabal, rather than just the mundane process of individuals going to their polling places and local election officials tallying their votes.

And conspiracy theories play into our powerful desire to belong to communities. In his book *How Minds Change: The Surprising Science of Belief, Opinion, and Persuasion*, journalist David McRaney considers people who believe that the Earth is flat. "At the end of the day," he writes, "it's the notion of a powerful *them* hiding the *truth* that motivated them to dig deeper, find one another, and form a community. They can all agree on that. Then group psychology took over; now they are bound by tribalism and reputation management, signaling to one another by committing to the central dogma that, no matter what, the Earth is not a globe."[28]

I'd like to think that the delusions of irrational thinkers will crumble if they just do some research. The catch, however, is that proponents of bogus theories love nothing more than proclaiming that they *have* done their "own research"—which all too often means that, spurred by confirmation bias, they just used the internet to find "alternative facts" that agree with their ideas. In 2024 a team of researchers from several U.S. universities examined if having participants do internet research would reduce belief in false facts. What they found was often the opposite. One of the authors reported that, "Our study shows that the act of searching online to evaluate news increases belief in highly popular misinformation—and by notable amounts."[29] One reason was that the participants tended to rely on information that had been boosted in search results. (A "fact" doesn't need to be accurate to reach the top; it just has to be frequently shared.)

As predicted by a cognitive bias called *the backfire effect*, even if you confront a flat-Earther with obvious proof that they're wrong (such as video filmed from a spacecraft), they'll often just double down. Why? If conspiracy theorists admit the validity of contrary evidence, they might lose their sense of themselves as shrewd and special. Worse, they might get ejected from their close-knit communities, and that's often too high a price to pay.[30]

In short, people can build strong communities around fallacies as easily as they can around truths—maybe *more* easily, since they often make extra efforts to defend a shared delusion.

It would be one thing if irrationality were limited to wacky conspiracy theories. The real problem is how it has infected our politics.

When a June, 2024 poll asked voters in six key U.S. swing states which presidential candidate would do a better job of handling threats to democracy in the U.S, eleven percent more of them picked Donald Trump over Joe Biden.[31] (This was after Trump spread his false claims about major fraud in the 2020 election; after he tried to strong-arm election officials to "find" him votes; and after he incited the Capitol insurrection.) It seems to me that a person would have to have ingested a massive amount of false or misleading information to reach such a bizarre conclusion, and I find it hard to believe that this result could have been possible if these respondents had been relying on a diet of news from traditional sources, such as mainstream national newspapers and network TV news programs.

And there's the rub. Vast numbers of them were not.

A stunning national April, 2024 NBC News poll[32] showed that of people who said they "Don't follow political news," or got their news from cable TV shows, 53% supported Donald Trump. 55% of people who got their news from national TV networks supported Joe Biden[k]— as did 70% of those who relied on newspapers.

I get my news from such heavily vetted sources as the *New York Times*, the *Washington Post*, CNN, and the *New Yorker*. The Gen Z-er who sent me the YouTube videos I mentioned earlier did so because she wanted to show me how people her age often get their political information from an entirely different, heavily digital mediasphere.

The first clip featured a 34-year-old "influencer"[l] sharing his thoughts on the video gaming platform Twitch. He offered a polemic titled "Legacy Media Realizes It's Dead,"[33] in which he argued that huge number of voters had abandoned mainstream media sources after realizing that they were extremely biased and full of lies. His "research" seemed to just consist of regurgitating clips from other influencers.

I'm always working to convince my students that they need to back up their arguments with evidence from reputable sources. For a professor who teaches journalism, it seems ludicrous that anyone should trust one lone video gamer who had clearly not done any actual reporting, nor undergone any review by editors or fact checkers over, say, the 3,347 veteran professionals who work at the *Washington Post*.[34] When I show my students the film *All the President's Men*, I'm always struck by how often *Post* executive editor Ben Bradlee told Woodward and Bernstein that he couldn't print their claims until they had been corroborated by multiple sources. *No one* was vetting the claims of the Twitch influencer. It might seem easy to dismiss him as just one uninformed amateur operating on "vibes," but here's the alarming thing: this cocky bro

---

[k] The poll was taken before Biden dropped out of the race.
[l] Someone—often without professional credentials—who has the ability to affect the opinions and behaviors of others, often through social media.

had more than *three million* YouTube subscribers, more than the entire print and digital subscriber base of the *Washington Post*.[35,m]

I turned to the other video, in which a 23-year-old conservative influencer did her best to convince her viewers that they had nothing to fear from Donald Trump—that he was essentially a nice, steady guy who had been falsely painted as extreme by liberal mainstream media.[36] I swear I'm not making this up: in the middle of listening to her argument, I paused to check the news, and learned that the President-Elect had just nominated congressman Matt Gaetz as Attorney General of the United States.[n]

This young influencer had four and a half million subscribers. Her video provided a perfect example of how out of touch with reality a "news" source can be when they don't do any actual reporting, yet a 2024 poll found that 37% of adults under 30 said that they regularly got "news" from such social media influencers —most of whom "have no affiliation or background with a news organization."[37]

The presidential election in that year was followed by an orgy of self-recrimination among Democrats and media professionals about what they had done wrong. In reality, many voters were just upset about inflation in consumer prices, which was actually a global problem attributable largely to a pandemic-caused disruption in supply chains.[38] But Democrats also faced another massive obstacle, which was that Big Tech's digital platforms such as Twitch streams, podcasts, and TikTok reels had enabled a radical bypassing of traditional media—and of fact checking. One of the two main presidential candidates repeatedly used his own social media platform to claim, for example, that undocumented immi-

---

[m] In November of 2024, it was 2.5 million.

[n] In other words, he had nominated someone for the country's top law enforcement post who was under investigation by his own Republican colleagues for sex trafficking of minors, illicit drug use, and other crimes.

grants were causing a horrific national crime wave, even though violent crime was actually significantly down from recent years.[39]

It would be easy to come up with twenty reasons that would help explain the 2024 U.S. election result, but it would be hard to deny the significance of this: Trump's campaign and political action committees shrewdly targeted his ads and appearances to voters who relied on digital streaming services for their "news," especially shows that featured influencers on the conservative "bro" end of the spectrum.[40]

Unfortunately for all of us, our elections might no longer be based on which candidates can come up with the best policies for the country, and then allow voters to make their choices based on accurate reporting—they might just come down to which party can do the most effective job of spreading false information, largely through digital media.º

The fact that influencers can reach millions of internet users doesn't prove that mainstream news sources are dead—or even that there's anything fundamentally wrong with them. It just shows that we need to support them, and help them compete with people who are extremely adept at using digital technologies to promote bad information.

After I watched the videos my young acquaintance sent me, we had a discussion about them. She told me that she doesn't trust any mainstream news organizations. I didn't try to argue with her—I just wanted to hear more about her beliefs. But when she claimed

---

º Donald Trump may well be the most prolific spouter of falsehoods in human history. The *Washington Post* Fact Checker team documented how he made 30,573 false or misleading claims in just his first four years as President. Their detailed database is staggering. (You can find it at https://www.washingtonpost. com/graphics/politics/trump-claims-database/?itid=lk_interstitial_manual_9)

that Trump had won in 2024 "by a landslide," I couldn't resist pointing out that he had won less than fifty percent of the popular vote.[41] She told me that she didn't believe this figure, because "it might be slanted by liberal bias."

How is it possible to maintain a democracy if citizens refuse to believe the actual tally of the votes?

Aside from doubting mainstream news media, she also said that she distrusted government agencies and scientists. If you deny all those sources of actual facts, it's a very short step to believing in conspiracy theories—and she said that she was inclined to consider a number of them, including the faked moon landing and the Bill Gates microchip fantasy.

Both conspiracy theorists and authoritarian leaders share a project: to get people to distrust experts and facts. Their motives are different, though. Conspiracy theorists just want people to believe their outlandish ideas, but autocrats want citizens to distrust public institutions such as the news media so they won't be believed when they call out the rulers' corruption and lies—or report who actually won elections.

Unlike conspiracy theorists, who genuinely believe their nutty notions, autocrats are deeply cynical: they lie intentionally in order to achieve domination and control. Their goal is to undermine a shared sense of reality, leaving them free to create their own "alternative facts."

The influencers in those two videos I saw seemed to genuinely believe that they were justified in encouraging mistrust of our society's fundamental institutions, yet they were unwittingly doing the despots' work.

It would be bad enough if social media companies just helped silo people into inwardly-focused and protective communities of false belief. Autocrats also often benefit from another age-old fact of life: tribes are often strengthened by creating false enemies from *outside* groups. Scapegoating, "the practice of singling out a person or group for unmerited blame and consequent negative treatment,"[42] has been used by bad actors throughout history, and it's often directed at those who are actually some of the most power*less* elements of society.

Despots know that they can distract their citizens from their own oppression if they give them the ability to punch down. The Nazis notoriously gave their movement cohesion by blaming Jews, Romani people, homosexuals, and disabled people for the ills of German society. Today, the powers-that-be have boosted their dominance by demonizing undocumented immigrants and trans-gendered people.

The propagation of an us-versus-others mentality may also help explain another bizarre mystery of our time: how can huge groups of people support politicians who egregiously violate those groups' professed moral codes? (During the 2024 presidential campaign, 82% of White evangelical Protestants supported Donald Trump.)[43]

One of the most cogent answers I have heard came from conservative *New York Times* columnist David French, who wrote about how "what we've been witnessing in the last decade is millions of Americans constructing a different moral superstructure,"[44] one in which a core sense of what is ethical and moral behavior gets subverted in favor of a code that says simply, *Whatever increases the power of my tribe is good, and anyone that decreases it is bad.*

"In the world of the friend-enemy distinction," French said, "your ultimate virtue is found in your willingness to fight. Your ultimate vice is betraying your side by refusing the call to political war."

This distinction doesn't just affect the way people vote; it affects how they govern when in power. The guiding calculus can change from what it should be—*What can I do that will benefit all of humanity?*—to *What can I do that will bring the greatest profits to my in-group?*

For those who think like that, major problems that we ought to be tackling, such as the climate crisis, economic inequality, and global poverty, often don't even figure into their concerns.

As harsh as the social media-enabled scapegoating of vulnerable populations has been in the U.S., it has often been even more extreme overseas, including horrific acts of political and ethnic violence. In Myanmar in 2018, for example, government officials used Facebook to spur a genocide of the Rohingya people,[45] and in that same year a political party in India used WhatsApp to spread rumors that its opponents were abducting children and harvesting their organs, which incited followers to commit a series of lynchings.[46]

In a 2020 internal memo, data scientist Sophie Zhang told her bosses that, "In the three years I've spent at Facebook, I've found multiple blatant attempts by foreign national governments to abuse our platform on vast scales to mislead their own citizenry." She reported evidence of massive campaigns to create thousands of fake accounts and influence elections in places like Honduras, Bolivia, Ecuador, and Azerbajian. Zhang says she was fired for "poor performance"[47] when she refused to stop focusing on such problems.

Her sad conclusion about her own culpability in working for this deeply flawed digital platform?

"I have blood on my hands."[48]

Katy Cook points to evidence of a major decline in democracy around the world,[49] and notes that it seems to have started in 2006 and 2007. Those years were marked by the launch of Twitter and Facebook,[50] Google's acquisition of YouTube, and the release of the first iPhones.[51]

There are many things we could do to ameliorate these political effects of digital technologies, and we'll examine them in the latter part of this book. For now, I'll just point out that we could require tech companies to do more content moderation; to change their algorithms so they don't push users toward extreme ideas; to apply a little friction to the sharing of posts (so that bad info can't go viral so quickly); and to take more responsibility for the content they disseminate.

After 2016, when the negative effects of social media became more publicly apparent, companies such as Facebook and Twitter were shamed into devoting more resources to dealing with the problems, but they had obviously not been fixed by January 6, 2021, when disinformation-inspired users committed their insurrection inside the U.S. Capitol. In response, the platforms finally banned some of their most incendiary posters, and it seemed that they would take other measures to stem the tide.

Since then, amazingly, the trend has gone in the opposite direction. As part of his disastrous takeover of Twitter, Elon Musk allowed people who had been banned for spreading disinformation to return to the platform, and he led the way for other social media companies to cut back on their own efforts to police misleading claims.[52] They gutted their content moderation teams, and reduced efforts to fact check and label false information. The lie of a rigged 2020 election was allowed free rein again, and hate speech flared up once more—all during the run-up to the 2024 elections.

Prior to that year, owners of social media platforms often

claimed that they were merely providing neutral information high-ways, and thus were not responsible for bad information posted by users. Musk dropped that mask of impartiality when he bought Twitter, renamed it X, and soon began using it to spread a torrent of his own bad information. One study found that his false or misleading claims had gotten two billion views in the run-up to the voting, leading Reuters to headline an article "Musk and X Are Epicenter of US Election Misinformation, Experts Say."[53] And those were just claims about the election; they didn't include his firehose of misleading posts or links about other topics, ranging from Covid misinformation to antisemitic conspiracy theories.[54]

Two months after the 2024 election, Mark Zuckerberg announced that Meta would begin drastically reducing its fact checking and content moderation efforts. Joel Kaplan, his Chief Global Affairs Officer, said that, "we're going to start treating civic content from people and Pages you follow on Facebook more like any other content in your feed, and we will start ranking and showing you that content based on explicit signals (for example, liking a piece of content) and implicit signals (like viewing posts) that help us predict what's meaningful to people."[55] In other words, the platform would return to letting its algorithms push partisan political content, and foster bubbles. It also curtailed its policies on "Hateful Conduct." (To give just two examples, it would now allow "allegations of mental illness or abnormality when based on gender or sexual orientation,"[56] and allow users to refer to women "as household objects or property."[57])

It wasn't clear if these moves were just a cynical ploy on Zuckerberg's part to curry favor with the incoming president,[p] or

---

[p] In 2021, veteran tech journalist Walt Mossberg said this about the tech titan: "A long time ago, I was very close to Steve Jobs and Bill Gates. [...]. They could be very unpleasant, but they had some principles. They had a red line. In my encounters with Mark Zuckerberg, I've never been able to discover any principles." He also said that, "I have never detected a moment when [Facebook] gave up profits to do

an indication of an increasingly right-wing bent on his part, but in light of what we've examined here, it seems likely that the political and personal consequences will be dire.

In 2025, YouTube privately loosened its moderation guidelines. As a *New York Times* sub-head put it, "The world's largest video platform has told content moderators to favor 'freedom of expression' over the risk of harm in deciding what to take down,"[58] re-opening the door to all sorts of false information and hate speech. As Imran Ahmed, the head of the Center for Countering Digital Hate put it, "What we're seeing is a rapid race to the bottom. The changes benefit the companies by reducing the costs of content moderation, while keeping more content online for user engagement." He added that, "This is not about free speech. It's about advertising, amplification, and ultimately profits."[59]

To sum up these past two chapters, social media platforms have had an unprecedented impact on our psychological and political lives. As Roger McNamee puts it, "You marry the social triggers to personalized content on a device that most people check on their way to pee in the morning and as the last thing they do before they turn the light out at night. You literally have a persuasion engine unlike any created in history."[60]

The consequences have been devastating. According to Jaron Lanier, "It makes everybody more and more vain, paranoid, irritable, xenophobic, stupid, fearful. And then we lose the ability to talk to each other."[61]

Autocrats around the world are now busy employing digital surveillance,[62] data collection, and dissemination of false informa-

---

the right thing for people." (https://www.nytimes.com/2021/10/18/opinion/sway-kara-swisher-walt-mossberg.html?showTranscript=1)

tion through social media as tools of political manipulation and repression, spurring frequent references to Big Brother in George Orwell's *1984*. We should note, however, that many of the technologies that make these things possible were not invented by governments, and that the Big Tech companies that created them have different goals. Shoshana Zuboff makes a distinction between authoritarian power and what she calls *instrumentarian* power. Unlike a totalitarian government, which is hugely concerned with controlling its citizens' political behavior and even their thoughts, the wielders of this new kind of sovereignty are "profoundly and infinitely indifferent to our meanings and motives. Trained on measurable action, it only cares that whatever we do is *accessible* to its ever-evolving operations of rendition, calculation, modification, monetization, and control."[63]

In short, the technologists who developed the internet and social media didn't set out to damage the mental health of their users, distort our sense of truth and reality, and jeopardize democracies all over the world.

Some just wanted to get rich by selling ads.

Chapter 4
What If Different
Choices Had Been Made?

WHEN I TEACH THESIS WRITING TO GRADUATE STUDENTS, I'M dismayed to come across sentences like this one: *The Internet appeared in the early 1990s and it transformed the world.*

I emphasize the fact that, throughout the entire history of the world, no technology has ever just "appeared" (as though, one day, it fell from the sky). It takes *people* making *decisions* to create one. **Specific individuals at Facebook and Google and the other big tech companies made specific choices that contributed to the problems we've just examined.**

If we think of technologies as independently arising, we can feel helpless to challenge them. On the contrary, if we see how people are creating them, we can protest if they seem to be leading toward bad consequences. And we can demand more say in how—or *if*—they're developed in the first place.

Many of the creators of the technologies we've just looked at have been dismayed by the consequences of their own decisions—and

I'm not just referring to our digital addictions and the political state of the world.

In contrast to the early notion that the internet would enable a remarkable, freeing decentralization of power, spread among a vast wonderland of little personal websites, by 2022 just Google, Facebook, Netflix, Amazon, Microsoft, and Apple generated nearly 57% of global network traffic.[1]

A few independent originators of content—writers and musicians and the dreaded influencers—earn money via the Web, but many have seen profits from their work flow into the pockets of a small number of content aggregators and streaming services and social media companies. We have shifted from a view of computer users as members of a network of independent creators and sharers toward one in which we're just sources of income for huge corporations.[a]

As for the ways in which the once commerce-free internet has been taken over by businesspeople intent on exploiting user data, things didn't have to turn out this way. Shoshana Zuboff notes that the surveillance capitalism model "is not an inherent result of digital technology, nor is it a necessary expression of information capitalism. It was intentionally constructed at a moment in history."[2]

What characterized that moment? As we have seen, government support of tech companies enabled engineers to lay the technical groundwork, and venture capital provided much of the rest of the funding. In the 1970s, neoliberal economists and politicians gutted anti-trust laws and deregulated markets, setting the stage for the advent of Big Tech. But few decisions shaped the era as much as a single line of text tucked into a 1996 law.

---

[a] To be fair, today's users *do* contribute a great amount of content in such forms as tweets, TikTok clips, and reader comments and reviews—but how much of it provides a substantial contribution to our culture and society? And how much of it earns money for the people who create it?

Section 230 of Title 47 of the U.S. Code, crafted by U.S. Senator Ron Wyden and Representative Chris Cox, was ostensibly designed to ensure that internet-based companies could screen out offensive material provided by their users, but it ended up having the opposite effect.

One clause stated that *No provider or user of an interactive computer service shall be treated as the publisher or speaker of any information provided by another information content provider.*[3] Since the passage of the law, that sentence has protected social media companies from legal liability for what third parties post on their websites.

The Electronic Frontier Foundation, a big booster of Section 230, argues that it "has allowed the internet to thrive."[4] Unfortunately, as we'll see in a moment, that little sentence has also allowed all sorts of harmful material to flourish on the Web.

The emergence of so many problems begs an essential question: what if the people involved in the invention of these technologies had made different decisions, and designed their software and business models in other ways?

Steven Johnson has made several arguments in this regard. He believes that a decision to use better online cryptography might have spared internet users from email hacks and identity theft; that different means of funding companies could have encouraged them to keep their software open, rather than closed; and that we could have designed a secure open standard to establish user identities on the Web, rather than moving in the direction of anonymity.[5]

The latter issue is a major one, and it was decided early on.

The designers of the internet set it up so that every computer connected to it bore a fixed IP address.[b] Message boards, in which people could gather online to chat about whatever they wanted, were one of the most popular early uses of the network. But if someone logged in on their work computer and wanted to chat about a sensitive topic (such as why they disliked their boss), they'd be taking a big risk if they posted under their real identity. The moderators soon decided to let users employ pseudonyms.

One of the best-known early online forums, launched in 1985, was called The Well (short for "Whole Earth 'Lectronic Link"). Over time, its civil, happy vibe was threatened as anonymous commenters began to inject vitriol and snark into the dialogues, and "flame wars" heated up.[6]

I saw examples of such bad behavior when I wrote an article for the *New York Times* about online neighborhood blogs. I was impressed by how readers used the Comments sections to share local news and debate community issues, but several moderators told me how they struggled to protect their sites from malicious anonymous commenters. One recalled having to deal with a troll who loved to shock other users by posting things like photos of severed heads. Another told me that, "I believe that anonymity, while freeing people to talk, well, freely, also causes this odd dissociation that leads them to say things they'd never dream of saying in real life."[7,c] The programmers who enabled anonymous posting failed to predict how it would contribute to a harshening of speech all over the Web.

To give credit where it's due, one good decision that Facebook made early on was to require users to post using their real identities. Unfortunately, people soon found ways to create fake accounts so they could get around that requirement, and some

---

[b] Internet Protocol

[c] Psychologists such as John Suler have called this "the Online Disinhibition Effect." (https://doi.org/10.1089/1094931041291295)

created accounts for "users" that were actually digital bots.[d] Some profiles were created by scammers who used them to con people out of money; others were used to falsely boost the number of fans and Likes on social media pages; and foreign governments created some to sow political disinformation. These problems grew so severe that Facebook announced that it had removed 2.2 billion fake accounts in just the first quarter of 2019.[8]

Some have argued that internet anonymity can be beneficial, as when it protects people who might be subject to abuse and discrimination,[9] and that's clearly sometimes true, yet I would guess that—on the whole—it has enabled more abuse than it prevented.

Providing the ability for users to act anonymously can abet the propagation of malicious content, making it another element of our Wheel.

The bad effects of the decision to allow anonymity became supercharged when combined with the consequences of a momentous decision we looked at several pages ago: the passage of Section 230.

Gilad Edelman, a writer for *Wired* magazine, notes the paradox contained in that hugely consequential piece of legislation: "The law was supposed to encourage online service providers to police their platforms without fear. It was, after all, part of a statute called the Communications *Decency* Act. And yet the first part of the law [...] removed a major legal incentive for them to police their platforms *at all*. With a few exceptions (most notably copyright infringement and child pornography), providers would

---

[d] Software programs that imitate the behavior of humans, as in participating in online discussions.

never be held responsible for material posted by users, no matter how clearly false or harmful. Even if they were put on notice, could easily fix the problem, and simply chose not to."[10]

As former Facebook Platform Operations Manager Sandy Parakilas has noted, "In terms of design, companies like Facebook and Twitter have not prioritized features that would protect people against the most malicious cases of abuse. That's in part because they have no liability when something goes wrong."[11]

Since its enactment, Section 230 has been used in court to defend all sorts of bad behavior, including an ad falsely associating a private citizen with sales of T-shirts mocking the victims of Timothy McVeigh's terrorist bombing in Oklahoma City, and posts accusing an NFL cheerleader of sleeping with her entire team.[12] Section 230 has even been invoked to indemnify companies for posting ads for illegal gun sales,[13] adult prostitution, and sex with trafficked children.[14]

Wheel factor: laws that deny corporate responsibility and liability.

Social media companies argue that they are merely platforms providing a forum for the expression of "free speech." That term is often invoked as an absolute, blanket right, protected by the First Amendment, but this reflects a major misunderstanding. The amendment states only that "*Congress* shall make no law [...] abridging the freedom of speech"—it doesn't say that private companies aren't free to moderate their own content.

Proponents of Section 230 argue that revoking it would have catastrophic effects for a wide range of companies across the internet because they'd be forced to take legal responsibility for, and monitor, every bit of user-created content that might appear on their websites. As an alternative, in the last section of this book, we'll look at ways in which it might be constructively amended.[e]

---

[e] Oddly, Section 230 has come under recent attack from both sides of the U.S.

Here's another decision point: the original Web protocols developed by people like Tim Berners-Lee didn't track us online. That ability was added by engineers in private companies, in the form of tools like cookies.[15]

And there were alternatives to the commercialization of the Web through advertising, as we might have learned from a look back to the early days of radio. In 1927, the British parliament established the British Broadcasting Corporation, or BBC, a public service which was set up to be funded by the national government —and it still is.[16] In the U.S. we went the way of paid commercials, despite Herbert Hoover's insistence that, "It is inconceivable that we should allow so great a possibility for service, for news, for entertainment, for education, and for vital commercial purposes to be drowned in advertising chatter."[17]

How might the internet look today if we had maintained the ban on commercial uses, or at least not let advertising become the main economic driver? And how might things have been different if venture capital had not become the major investment source for new companies—if we had found other ways to support fledgling innovators? What if engineers had never invented features designed to addict us to social media, and algorithms that polarize us and push us in extreme directions?

I don't think it's hyperbole to say that we might be living in a better world.

Of course, certain choices *were* made. Surprisingly, some have been deeply regretted by the very people who made them. In a stunning article titled "The Internet Apologizes for Violating and

---

political aisle: from Democrats who say that social media companies should have to take more responsibility for the hate speech, misinformation, and lies that flood their sites, but also from Republicans, who claim that the companies have used it to block content from conservatives.

Hijacking Our Attention," journalist Noah Kulwin included rueful quotes from Guillaume Chaslot, the engineer who designed YouTube's algorithmic recommendation system, and Ethan Zuckerman, the inventor of the pop-up ad, who said that, "I really did not mean to bring this horrible thing into people's lives."[18] Other tech workers who have lamented features they unleashed upon the world include Justin Rosenstein, the inventor of the Facebook *Like* button, and Loren Brichther, who devised the addictive pull-to-refresh capability on Twitter feeds.[19]

It's clear from a number of insider accounts that decisions like those were made because companies like Google and Facebook and YouTube were focused on optimizing their bottom lines. As Dan McComas, former Senior Vice President for Product at Reddit, put it, "The incentive structure is simply growth at all costs. There was never, in any board meeting that I have ever attended, a conversation about the users, about things that were going on that were bad, about potential dangers, about decisions that might affect potential dangers."[20]

Ultimately, the lesson is clear, and again, this is part of the essential premise of this book: **If tech leaders would recognize the factors that make bad consequences more likely to arise, they could do a better job of predicting them, and that could help them make better choices *before* things go wrong.**

Chapter 5

"The Sword And The Spear": The Danger Of Technological Arms Races

IN A 2015 SURVEY, 71% OF THE AMERICANS POLLED SAID that tech companies had a positive effect on the country.[1]

By 2021, that number had plummeted to 34%.[2]

That period of *techlash* was a particularly rough one for Facebook's reputation, especially after the eruption of the Cambridge Analytica scandal in 2018[3,a] and Frances Haugen's whistleblowing in 2021. As he is wont to do, Mark Zuckerberg offered seemingly contrite public apologies.[4]

It seemed like a time of reckoning, and it could have marked a major positive turning point, if Zuckerberg and other Silicon Valley leaders had reflected on how their design and engineering decisions had led to damaging consequences, and made a bigger effort to fix the flaws. It could also have been an opportunity to change how they were designing newer technologies, to slow down and ensure that they were building them in more prudent ways.

---

[a]   The political consulting firm acquired the private personal data of tens of millions of Facebook users in order to micro-target and manipulate swing voters in the lead-ups to the election of President Trump and the U.K.'s Brexit vote.

Instead, though—and it seems quaint to mention this now—a number of them were already plunging into developing another problematic social media platform. In his *Founder's Letter* of 2021, Mark Zuckerberg wrote that, "We believe the metaverse[b] can enable better social experiences than anything that exists today."[5] He was so sure that this new virtual reality social platform would become the hottest new form of tech that he even changed the name of his company to Meta.

Now, it's fair to say that such a platform might support all sorts of benevolent uses. It could, for example, offer useful simulators to help train surgeons and pilots, provide a semblance of mobility for severely disabled people, reduce the carbon footprint of physical commuting, and enable students to "visit" archaeological sites all over the world.

But the technology also seemed likely to push bad consequences. For one thing, since Big Tech engineers had already put a huge amount of effort into designing features to keep us glued to our little smartphone screens, could anyone imagine that an immersive 3-D environment in which we could look like anything we want and fly around like superheroes would somehow be *less* addictive? And in a world in which users would wear the cloak of anonymity provided by digital avatars, the trolling problems of today's social media would likely become even more extreme, as would kids' body image issues when surrounded by an endless supply of virtual supermodels and musclemen.

In an article titled "The Metaverse's Dark Side: Here Come Harassment and Assaults,"[6] reporters Sheera Frenkel and Kellen

---

[b] He spoke of a metaverse, but it might be more appropriate to talk of *metaverses*. (At this point, it's unclear if one company would have created a single one that would have dominated the field, or if various companies would have established their own competing virtual worlds. In 2023, the top companies investing in the technology included Microsoft, Google, Meta, Decentraland, Nvidia, Shopify, Unity Technologies, Roblox, and Epic Games.)

Browning described the experience of a female player of a virtual reality game called *Population One* in 2021. Another user approached her, and then "simulated groping and ejaculating onto her avatar. Shocked, she asked the player, whose avatar appeared male, to stop. He shrugged as if to say: 'I don't know what to tell you. It's the metaverse—I'll do what I want'." Meta claimed that it was developing safety measures to prevent such harassment, but its Chief Tech Officer John Bosworth became notorious for an internal company memo in which he admitted that moderating what people would say and how they would act in such a cyberspace "at any meaningful scale is practically impossible."[7]

Zuckerberg promised advertisers that the new platform would provide fantastic new ways to target consumers.[8,c] In the real world, we can at least look away from or mute ads, but they would probably be unavoidable on such a platform.[d] As for politics? The same virtual universe that would allow users to explore galaxies and float through oceans could also be used for holding fascist rallies or militia meetings.

It would seem that Mark Zuckerberg and the other proponents of the metaverse didn't learn much from their failures with social media. Given the revelations about the harms they had already caused, they could have abandoned the idea, or at least aimed for a more modest, controllable version, but the Wheel factors driving them proved far too strong. As of 2024, Zuckerberg had poured more than forty-five billion dollars into his company's metaverse-building Reality Labs.[9] For a while, the concept received tons of media attention as the Next Big Thing.

---

[c] He also expressed the belief that it would create a vast new marketplace for selling bespoke avatars, as well as digital clothing and goods. The sad thing is that there would undoubtedly have been users willing to pay for such nonsense.

[d] As former tech coder and startup founder Wendy Liu put it, "How would you explain the metaverse in 5 words or less? Best I got is 'virtual reality with unskippable ads.'" (https://twitter.com/dellsystem/status/1420818448456716289)

The reason it now seems quaint to mention it is that—despite Zuckerberg's best efforts—the idea soon curdled. Journalists began writing obituaries for it, such as this one from John Herrman in *New York* magazine:

> "Silicon Valley's brief obsession with the metaverse has assumed the quality of a bad dream, half-remembered. Legless avatars beckoned us into barren digital landscapes to . . . stand around and talk about NFTs? Parcels of "property" sold for millions of dollars? It was a . . . virtual world? No? A mixed-reality game? No? A new frontier? An escape from meatspace? A layer on top of it? Companies raised and spent billions of dollars on the metaverse without ever quite getting their stories straight about what it was supposed to be or do."[10]

In 2023, Microsoft, Disney, Walmart, and other companies that had invested heavily in metaverse projects began to lay off their engineering teams.[11]

What caused this unusual tech world about-face? The fuzziness of the concept was certainly one reason. Beyond a few obvious uses, such as gaming, entertainment, education, and the inevitable porn, the concept just didn't seem terribly *necessary*.[e] Few people seemed to share Mark Zuckerberg's enthusiasm for it, including many of the workers at his own company.[12]

But it was two other factors that really deflated the metaverse bubble.

One was a surprise development in public health: the Covid-19 pandemic. Many of the world's citizens were forced to spend several years staring at each other on computer screens, which

---

[e] I can't help but think about an old *Saturday Night Live* sketch in which executives at a personal care products company want a way to market a new range of cleansers and deodorizers. They invent a new body orifice, and they call it the *blowhole.*

hardly left us eager for more online interaction—especially not more virtual meetings (which Zuckerberg had promoted with stunningly awkward videos).[13] The fact that the metaverse would be accessed by clunky, expensive facemasks didn't help.

But the biggest reason for the drop in interest was another curve ball: in 2022, the whole concept was eclipsed by a suddenly viable new technology that offered an astounding range of practical uses, and promised venture capitalists an almost boundless return on investment.

That new Big Thing, of course, was—and is—artificial intelligence.[f]

By this point, so much has written about A.I. that you may well be tired of reading about it. I'm just going to mention how its development points to two more spokes in our Wheel, the second of which may be the most dangerous of all.

Proponents of the new technology have claimed that it might help us free humanity from drudge work, find cures for deadly diseases, improve weather forecasting, and conquer hunger, poverty, and global warming.[14] Those are certainly worthy goals, but it's also so risky that it could make the harms of social media seem like paper cuts. As I noted earlier, critics have warned that its development might have catastrophic unintended consequences. Some predict that human bad actors—authoritarians, terrorists, criminals—might use it to become enormously empowered. Others envision that an A.I. might become conscious and far smarter than humans, and maliciously decide to wipe us out. (Cue endless

---

[f]   I should point out that the death knells for the metaverse concept may yet prove premature, since work on building it is still proceeding at some companies—and advances in A.I. might actually make it more feasible.

references to *The Terminator*.) Others have pointed out that an artificial *super*intelligence—an *A.S.I.*— would not have to be malevolent, or even conscious, to wreak major havoc. Physicist Stephen Hawking argued that, "The real risk with A.I. isn't malice, but competence."[15] In other words, we might call on such a model to achieve all sorts of useful objectives, but it might accomplish them in disastrous, unexpected ways.

Swedish philosopher Nick Bostrom has been warning about the dangers of A.I. for more than two decades,[16] and illustrated the problem with a famous thought experiment. If an A.S.I. were given the instruction to make lots of paper clips, it might determine that it would be better able to achieve that task by getting rid of humans, since we might try switch it off before it could achieve that goal—or it might decide to break down all organic matter (including people) for atoms that it could use to make more clips.[17,g]

Does that sound fantastical and extreme? Let's try a more practical example: Wall Street financiers ask an A.S.I. to help them maximize the value of a corporation's stock, thinking that the model might perform an ultra-sophisticated analysis of investment strategies. Instead, the A.S.I. concludes—quite logically—that it can boost the stock by gaining access to the computer systems of rival corporations and destroying them with malware. (Perhaps destabilizing our economic system.)

The challenge of making sure that an A.I.'s methods are in sync with human needs for safety and stability is known at the *alignment problem*. Part of the solution might seem obvious: as science-fiction author Isaac Asimov suggested with his famous

---

g Human considerations of good and evil might be utterly irrelevant to such an ASI, which might only care about achieving practical objectives. Tech blogger Scott Alexander has described such an entity as "some kind of unfathomable blind idiot alien god." (https://slatestarcodex.com/2014/07/30/meditations-on-moloch/)

Three Laws of Robotics,[18] you just imbue your model with fundamental rules about not harming humans.

Unfortunately, devising foolproof rules turns out to be extremely challenging. When an interviewer asked Bostrom, "Could we make its primary goal be improving the human condition, advancing human values—making humans happy?" he replied, "Well, we'd have to define then what we mean by being happy. If we mean feeling pleasure, then perhaps the superintelligent AI would stick electrodes onto every human brain and stimulate our pleasure centers. Or you could take out the body altogether and have our brains bathing in a drug the AI could design. It turns out to be quite difficult to specify a goal of what we want in English—let alone in computer code."

I realize that the odds of a future computer superintelligence wiping out humanity might seem slim (though ten percent hardly seems slim enough). Even so, the technology presents a more immediate risk. In that major 2023 survey I referenced earlier, A.I. professionals estimated that their creations will be able to replace all human occupations within less than a century.[19,h] That might cause a massive job loss, which could spur economic, social, and political turmoil on a global scale.

Any way you look at advanced A.I., the hazards make it obvious that if we're going to develop it further, we should devote a huge amount of effort to solving its safety problems. Yet one

---

[h] Creative occupations are not exempt: these researchers estimated that there was a 50% chance that new A.I. models would be able to write a Top 40 pop song and a bestselling novel within ten years. Just two years later, the creators of a completely artificial rock "band" called The Velvet Sundown released an album that sounded so lifelike that it got more than a million (human) listens on Spotify. (https://www. berklee.edu/berklee-now/news/velvet-sundown-ai-band-controversy)

researcher estimated that in 2020, corporate investment in advancing A.I. capabilities was more than 2,000 times the amount spent on reducing risks.[20] By a 2023 estimate, the number of experts working full-time around the world to reduce existential risks was only around 400 people.[21]

Inadequate investment in safety programs is a major risk factor in our Wheel.

Speaking of safety, science fiction authors have long written about one essential measure that could protect us from an A.S.I. that might go out of control. As tech journalist Kelsey Piper puts it, "We would be way safer against A.I. takeover scenarios, really scary catastrophes, if A.I. systems didn't have open access to the internet and the ability to contact anybody they wanted and do anything they wanted on the internet."[22]

Sam Altman, the CEO of OpenAI, the company that released ChatGPT in 2022, which initiated the current boom, has acknowledged this danger, saying that, "An agent that had full control of the internet could have far more effect on the world than an agent that had full control of a sophisticated robot."[23] He added that, "The bad case—and I think this is important to say—is, like, lights out for all of us."[24]

This is the same Sam Altman whose company has already connected some of its most advanced models to the internet.[i]

"Of course, we've given them that," says Piper, "because it was slightly more profitable than not giving them that."

"As soon as you open the door for everybody to be competing with each other," she added, "then the people who are careful and cautious are slower than the people who are careless and incautious. And the people who are the most willing to throw safety

---

[i]   He has also allowed vast numbers of internet users around the world to connect to his A.I. models. "One hundred million people were using ChatGPT within weeks of its launch." (https://asteriskmag.com/issues/03/a-field-guide-to-ai-safety)

considerations out the window are the ones who are getting their biggest models out fast."

Originally, AI models were just connected to the internet for the purpose of gathering training data, but in early 2025, OpenAI unveiled a new tool called Operator that can engage with the Web more actively, doing things like making restaurant reservations or shopping for groceries. The company's product and engineering lead Yash Kumar said that, "It can navigate websites and take actions on websites, much like you and I do."[25]

Gee, what could possibly go wrong? Step by hasty step, AI company leaders are rushing to do exactly the things that dystopian science fiction authors have long warned us about. Things have gotten so fraught that, despite Silicon Valley's extreme allergy to government "interference," worried workers have actually begun asking for government controls. As journalist Ezra Klein put it, "Among the many unique experiences of reporting on A.I. is this: In a young industry flooded with hype and money, person after person tells me that they are desperate to be regulated, even if it slows them down. In fact, especially if it slows them down. What they tell me is obvious to anyone watching. Competition is forcing them to go too fast and cut too many corners. This technology is too important to be left to a race between Microsoft, Google, Meta and a few other firms. But no one company can slow down to a safe pace without risking irrelevancy."[26]

What Piper and Klein are describing is what I believe might be the most dangerous Wheel factor of all: **An arms race that pushes technologists to prioritize winning over every other consideration, especially safety.** (Sometimes these races are between companies; sometimes between countries.)

One obvious factor fueling the A.I. arms race is the fact that the technology offers the possibility of astounding financial rewards. As John Herrman puts it, it "represents, among other things, a profound tech-exec fantasy: an endless supply of cheap and obedient labor and a chance to take ownership of the means, of, well, everything."[27] If we're looking for reasons why today's A.I. entrepreneurs are rushing forward despite horrendous risks, this is certainly one of the main ones: their fear of the potential long-term dangers of the technology seems outweighed by their short-term fear of losing the chance to gain an unprecedented range of monopoly power.

Of course, technological arms races are hardly a new phenomenon. As Johns Hopkins political science professor Daniel Deudney has pointed out, "Across the history of warfare, every innovation has triggered a quest, successful in widely varying degrees, to find yet another weapon to counter or defend against it. Thus the sword and spear evoked shields and body armor, and aircraft stimulated the development of anti-aircraft guns and missiles."[28] Before we knew it, we had a planet bristling with nuclear weapons.

Developers of A.I. are fueling a new military arms race. If Isaac Asimov were alive today, he'd undoubtedly be bewildered and chagrined to learn that we humans *are actively designing human-harming robots*. Engineers in defense departments around the world are creating A.I.-guided machines that are not just meant to kill people, but to do so without being ordered to by other humans.[j]

Proponents of these weapons claim that they'll reduce the risk

---

[j] Their use is not just a remote possibility: it may have already happened. The *Washington Post* reported in 2022 that, "The hazards of AI weapons were brought home last year when a U.N. Security Council report said a Turkish drone, the Kargu-2, appeared to have fired fully autonomously in the long-running Libyan civil war—potentially marking the first time on this planet a human being died

of human errors that accidentally kill civilians, and they won't be misguided by human emotions, but a number of nongovernmental organizations have banded together in horror at the prospect of such "slaughterbots."[29] The Convention on Certain Conventional Weapons, a U.N. gathering that meets every five years, has tried to create a treaty that would ban these Lethal Autonomous Weapons Systems, but it has been stymied because a number of its members, including the U.S., China, and Russia, are all racing to develop them first.[30] C. Anthony Pfaff, a retired Army colonel and a professor at U.S. Army War College, has argued that, "Any effort to ban these things is futile — they convey too much of an advantage for states to agree to that."[31]

Another technological arms race, which has been largely overshadowed by the one for more powerful A.I., is taking place in quantum computing. This technology may soon allow us to build computers so incredibly fast and powerful that they would make today's best devices seem like crude children's toys. They could bring all sorts of benefits, enabling major advances in areas such as medicine, chemistry, and green energy. One huge problem, however, is that they might also make it extremely easy to crack our current systems for encrypting information—which could wreak havoc on all sorts of systems, including credit cards, banking, air traffic, stock exchanges, nuclear power plant security, and military intelligence.[32,k]

China, Russia, and the U.S. are rushing to be the first country to create a useful quantum computer, which could give it enormous financial and military advantages over the others—and this has set off a secondary frantic race: to develop stronger encryption before one nation can come out ahead.[33]

---

entirely because a machine thought they should." https://www.washingtonpost.com/technology/2022/03/11/autonomous-weapons-geneva-un/

[k] Whereas it might take conventional processors thousands of years to break today's security codes, a quantum device might do so in *minutes*.

The problem with such contests is that they can resemble tsunamis: they have enormous momentum, and are extremely difficult to stop.

But tsunamis are out of human control, and technologies are not.

Every once in a while, we have already managed to band together to avoid entering these deadly races, as when world leaders signed treaties to outlaw chemical and biological weapons, to restrict the propagation of nuclear weapons, and to limit the militarization of outer space. We have realized that no nation can win if the whole world ends up losing, and we can do so again.

Part Two

CAUSES

<u>**On the Culture, Psychology, Ideology, Politics, and Economics of Silicon Valley**</u>

I started out by asking why people who have warned about the existential dangers of artificial intelligence might still plunge ahead to develop it. The arms race dynamic goes some way toward explaining this paradox, but there's a lot more to the answer. Today's A.I. story is emblematic of broader inclinations of tech leaders to ignore risks, neglect safety, and prioritize power and profit over more human and humane concerns.

Unfortunately for all of us, those tendencies are encouraged by a number of conditions within Silicon Valley.

In this section, we'll examine its culture, psychology, politics, and ideology, and the economic system that provides its beating heart.

Chapter 6
"The Technical Problem Captivated Me": On Psychology And Culture

LOSING ONE'S JOB CAN BE A DEVASTATING EXPERIENCE. I REmember a heartbreaking spate of suicides in New York City in 2018 by taxi and car service drivers who were distraught at how Uber and other new digital ride-hailing services had undermined their ability to make a living.[1]

Now think about how much greater that pain might be if we make self-driving vehicles the norm, and all the people around the world who drive professionally—including all the Uber drivers— find that not only their personal jobs, but their whole employment field has been wiped out. There's a word for that kind of imagination: *empathy,* which is a major component of what's known as *emotional intelligence.* It requires the ability to be aware of the thoughts and feelings of other people, and to imagine what it might be like to be in their place.

A tech creator with strong emotional intelligence might think about all the people who could be put out of work by self-driving vehicles, and consider whether the benefit to society would be worth the pain. Who knows? Maybe they'd decide that it *would* be worth it, because they'd be most concerned for the 1.35 million people who die—often due to human error—in road accidents each year.[2] But they might also consider how to ease the social impact of the transition. Whenever I read about the people who are designing driverless vehicles, though, I'm struck by how rarely (if ever) they mention the millions of workers their inventions might displace. They just seem fixated on solving the technical problems, and winning their arms race.

Tech journalist and venture capitalist Om Malik has argued that "Silicon Valley's biggest failing is [...] the distinct lack of empathy for those whose lives are disturbed by its technological wizardry."[3]

What's behind this shortage of concern? To understand it, it might be helpful to take a look at the broader culture in which today's technologists operate.

Consider the story of one lucky young man. While he was enrolled at Stanford University at the turn of the millennium, Ezra Callahan sold advertising space in the student newspaper to local businesses. It was a mundane work experience, but within a decade it would have a stunning payoff.

When he graduated in 2003, Callahan moved into a house near campus with four male college buddies. They had an extra bedroom, which they rented out for a short time to a 24-year-old named Sean Parker,[4] who had made quite a stir in the tech world four years earlier when he co-founded Napster, the notorious music file-sharing service.

In 2004, Parker became the first president of a tiny company founded by several former Harvard students he had recently met. The fledgling company wanted to make some money through ad sales, so Parker thought his recent roommate might be able to help, and Callahan was hired as its sixth employee. When he left in

2012, he owned what might have seemed a very modest stake in the company's shares—except the company's name was Facebook, and his .08% was worth $68 million.[5]

Callahan later looked back on his experience and wondered, "How much was the direction of the internet influenced by the perspective of nineteen-, twenty-, twenty-one-year-old well-off white boys?"[6]

In 2010, anthropologist Joseph Henrich and psychologists Steven Heine and Ara Norenzayan argued[7] that research studies in the field of psychology tended to be skewed because most of the subjects were college undergraduates, and they often had traits that were not universal. They were likely to be members of groups that were Western, educated, industrialized, rich, and democratic, a quintet of qualities the authors combined in the acronym W.E.I.R.D. Most professionals in Silicon Valley fall within most of the parameters of that set, but they also comprise an even smaller subset: most of the above, plus young and white. And like Ezra Callahan, they tend to make advantageous connections while attending a small number of feeder colleges.[a]

Adrian Daub remarks that, "Silicon Valley loves the words 'everyone,' 'universal,' and 'people,' but what they usually mean is 'people I went to school with,' 'my housemates in Palo Alto,' or 'my four immediate subordinates.' The universality that their business model pushes them toward exists in tension with the fact that they actually know very few people."[8] As the story of Ezra Callahan would indicate, however, things can work out very well for them if those few friends happen to be at the centers of power.

Things don't work out as well for people who don't fit the Big Tech mold. Although the 2010 census showed that 16% of Ameri-

---

[a] A 2018 survey of venture capitalists in the U.S. found that 40 percent of them attended Stanford or Harvard. (https://medium.com/@kerby/where-did-you-go-to-school-bde54d846188)

cans were Hispanic and 13% were Black,[9] a 2018 study[10] found that in Silicon Valley, the percentages of professional tech jobs held by Hispanic and Black workers were at 5% and 2.9%, respectively.[b] When it came to managerial or executive positions, the percentages shrank even further.[c]

While young white graduates of a few universities dominate Silicon Valley, it's only a subset of this subset that has gained enormous wealth and power in the last several decades. According to Paulina Borsook, they share a certain profile. "The entrepreneur personality—which needs little downtime, which must be narrowly focused and not prone to self-doubt, which will do all and anything to succeed, which tirelessly and compulsively must act like the greatest salesman in the world, which by definition is workaholic, which risks (and maybe devalues) family life and health—thrives in this ecosystem."[11]

Clearly, that's a difficult environment for young parents or those with other family obligations outside of work, and that's just one reason why the subset narrows even further: it's overwhelmingly male. By 2020, women made up a slight majority of the U.S. population, but they held only a quarter of professional tech jobs. When it came to black women, that dropped to 2.2%, and it was even lower, at 1.7%, for Hispanic women.[12] Startups led by women in 2020 received only 2.3% of venture capital funding[13]— and that figure was trending down.

The low percentage of females employed in Silicon Valley is obviously bad news for women, but it's also not good for that

---

[b] Other tech centers are considerably more diverse. The percentage of Black tech workers in the Washington D.C. metro area, for example, was over 17 percent in 2017.(https://www.theguardian.com/technology/2017/aug/07/silicon-valley-google-diversity-black-women-workers)

[c] In contrast, South and East Asians fared much better in Silicon Valley than in their overall U.S. percentage of 6%, and there are now a few Asian top executives, including Google CEO Sundar Pichai and Microsoft CEO Satya Nadella.

culture as a whole. Psychologist Katy Cook, the author of an excellent book titled *The Psychology of Silicon Valley*, cites researchers who have shown that "qualities such as collaboration, empathy, open-mindedness and maturity, and social and emotional skills tend to be more prevalent amongst women than men."[14]

This shortage of women in the industry can affect the technologies its workers create. For example, Cook notes that, "By assigning female rather than male voices and personas to popular digital assistants such as Amazon's Alexa, Apple's Siri, Microsoft's Cortana, Google's OK Google, and Facebook's now defunct M, there is an implicit correlation between women and helping and administrative positions."[15]

Proponents of A.I. argue that its algorithms can avoid the cognitive biases and prejudices that we mere *homo sapiens* are so prone to embody. For example, human judges have been shown to be more lenient on defendants if they have just eaten lunch.[16] That's not a problem a computer program is going to have. But it's not easy to get away from human inequities. In an article titled "Artificial Intelligence Has a 'Sea of Dudes' Problem,"[d] journalist Jack Clark argued that, "If everyone teaching computers to act like humans are men, then the machines will have a view of the world that's narrow by default and, through the curation of data sets, possibly biased."[17]

Amazon unintentionally provided a great example of this in 2014, when it decided to streamline its hiring process by using an early version of A.I. to sort through the many thousands of job applications it received every year. It seemed logical to think that the process should scan for keywords that had appeared on the resumés of past company hires. Soon, though, a big problem emerged: the new system seemed prejudiced against women. In fact, it downgraded resumés that merely mentioned the word, as in

---

[d] Clark attributed the term to Margaret Mitchell, a researcher at Microsoft.

"women's soccer captain" or "women in business." The program had been trained how to operate by analyzing data, and that data was riddled with biases. Faced with the evidence, the company abandoned the new tool, and returned to letting humans decide how to hire other humans.[18,e,f]

In 2020 Timnit Gebru, then the co-lead of Google's Ethical Artificial Intelligence Team, co-authored a paper with five other researchers[19] that explained why the training data for Silicon Valley's large language models (the basis for A.I. chatbots) tends to skew in the direction of certain worldviews: it includes mostly information written in English; it overrepresents data from younger users and those from developed countries; and it relies heavily on sources such as Reddit and Twitter, which overrepresent extreme viewpoints. The authors suggest that it would be better to carefully curate training sets, rather than just "ingesting massive amounts of data from convenient or easily scraped Internet sources"—which is how most data sets are built now.

It's not just particular biases that get passed on in tech, but the nature of the culture itself. As Katy Cook explains, "One dynamic

---

[e] Tech researcher Kate Crawford notes that "examples of discriminatory AI systems are legion, from gender bias in Apple's creditworthiness algorithms to racism in the COMPAS criminal risk assessment software to age bias in Facebook's ad targeting. Image recognition tools miscategorize Black faces, chatbots adopt racist and misogynistic language, voice recognition software fails to recognize female-sounding voices, and social media platforms show more highly paid job advertisements to men than to women." (*Atlas of AI*, p. 128)

[f] To be fair, some companies are working to counter such biases, but these efforts have sometimes backfired in interesting ways, as when, in 2023, Google paused its Gemini model after it was shown to be doings things like generating images of Asian and Black people when prompted to create an image of "a 1943 German soldier," or images of "the U.S. Founding Fathers." (https://www.theverge.com/ 2024/2/22/24079876/google-gemini-ai-photos-people-pause)

that perpetuates the homogeny of the industry is what companies in Silicon Valley refer to as 'culture fit,' which is the idea that to be a good addition to the organization, you must possess the same qualities as those already employed within it. [...] The result is an industry that has a great deal in common with itself and is comprised primarily of people with similar backgrounds, perspectives, and values. The idea of culture fit is so deeply embedded within the vocabulary of Silicon Valley that Google famously has its own word for it: Googley."[20]

In effect, the Valley is a very powerful filter and preference bubble.

To fit into it, it certainly helps if you're young.[g] Younger workers enjoy advantages such as the deep familiarity and ease with tech that comes from being a digital native, and an openness to innovation, aided by a lack of attachment to conventional ways of doing things. An important asset that can come with age, though, is a perspective on the history of one's company and of the industry, with accumulated memories of what caused problems in the past. But here's how Sandy Parakilas described Facebook's workplace culture: "I would argue that there's really no sense of history. It was more of a sense that we are building the future and there was a real focus on youth being a good thing. [...] No one understood that things might go badly."[21]

Technology has long been a field in which people can find success at an early age: a great invention is a great invention, regardless of the age of its inventor. (Thomas Edison, for example, received his

---

[g] Although the median age of U.S. workers in 2021 was 42, the median at Facebook was 28, for example, and 29 at SpaceX and LinkedIn. (https://www.bls.gov/emp/tables/median-age-labor-force.htm)

first patent at 22.[22]) What makes Silicon Valley unusual, though, is that—unlike the traditional corporate world, in which, generally, only seasoned, mature managers can rise to the top—clever tech entrepreneurs have been able to found and lead their own huge companies at stunningly early ages, without having to climb any corporate ladders at all. (Mark Zuckerberg started Facebook when he was just 19, and had become a billionaire by the time he reached 23.[23])

Unfortunately, the aptitudes required to pull off a feat like that don't necessarily include emotional intelligence. When I interviewed Katy Cook, who has spent a great deal of time working with people in the world of tech, she drew a connection between maturity and empathy. "I think age provides a lot of wisdom, but also humility, the ability to listen and self-reflect, and all these things that we associate with better emotional intelligence. Because emotional intelligence is a set of skills. And you usually learn those as you go along in life."[24]

"For me," Cook said, "the biggest defining factor between the healthier companies and the less pro-social companies was usually the age of founders. From my perspective, a lot of young founders are very ill-equipped socially and emotionally to handle that role. And that then bleeds into the culture."

Immaturity often leads to poor judgment, which makes it another spoke in our Wheel.

To be clear, I'm not trying to draw an inevitable connection between youth and immaturity. I see students on campus every day who are earnestly concerned about things like global warming and environmental sustainability, and are very interested to figure out how they can help make the world a better place. But anyone who has followed the history of Silicon Valley in the past three decades has seen an odd combination of traits in too many of its leaders: a technical and entrepreneurial brilliance paired with an emotional callowness, whether we're looking at the fact that Mark

Zuckerberg's prototype for Facebook was designed to show users photos of two female Harvard students and ask them to vote on which was hotter,[25] or the rampantly sexist bro culture Travis Kalanick established at Uber before he was forced to resign,[26] or the fact that Elon Musk has shown an odd fascination with the poop emoji.[27,h]

It's clear that if these CEOs comport themselves badly, that can have serious financial consequences for their backers. "So why" asked *Providence Journal* columnist Froma Harrop, "do capitalists put enormous valuations on companies run with all the social maturity of an unruly frat house? [...] The answer may be that many of the investors are themselves bros. They admire those overly confident rule breakers and share their indifference to ethical concerns. Where others see arrested development, they see disrupters. Where others see repellent behavior, they see tough entrepreneurs."[28]

Professor Daniel Deudney (who we'll meet in an upcoming chapter) has done a great job of summing up this kind of mindset in the world of tech. "Impulses still outrun prudent calculation of risks. Impatience is great and attention spans are short; no one wants to make their bed or take out the garbage. Restraints chafe; no one wants to close off options or forgo possibilities. Interdependence seems threatening to freedom. Responsibility is dreary and must be evaded. Hopes and dreams seem more real than dangers and problems."[29]

This strain of immaturity has played out in a broad resistance to oversight and regulation in Silicon Valley, with the government cast as a resented parent. And it's apparent in the way some devel-

---

h   One would hope that Zuckerberg would have grown more mature with age, yet in 2023 he accepted Musk's challenge to a public cage-match fight. (Though, like many adolescent boys, they didn't go through with it.)

opers have insisted on creating A.I. chatbots free of the safety controls that other companies have imposed.[30]

Let's take a moment to add several spokes to our Wheel. An overly homogenous, insular culture can foster discrimination within that culture, lead to a lack of concern for people who don't fall inside its parameters, and contribute to a shortage of empathy and other forms of emotional intelligence.

A culture that lacks diversity in age, and that tolerates immature behavior, will also be deficient in that regard.[i]

A shortage of emotional intelligence can make it harder to care about, predict, and prevent negative consequences.

I don't want to tar everyone in Silicon Valley with this brush, because only a portion of the professionals there fit the profile. (Too often, they're ones in leadership roles). For every immature founder with visions of dominating their industry, there are plenty of brilliant workers who just love solving problems, writing good programs, and coming up with elegant designs.

Former coder and tech executive Robert Gardiner told me that,

> "For me, a lot of it was interesting and challenging problems to solve, and then when my partners and I had our company there was also the excitement of building something from nothing. That is very heady, to see this little thing that started with five

---

[i] A culture that skews too old may have its own problems. (See for example, today's U.S. Congress.)

people and now it's twenty-five people and now it's fifty and then a hundred people. For me, a big part of it was the people that worked for us, the opportunity to give people—especially right out of college—interesting work and responsibility and train them and give them skills, that was awesome. And it's an amazing outlet for creativity, and at the end of the day, before you go home, to look at what you built that day, to look at the lines of code that you wrote, this program you created, these new functions that you added or whatever—it's really great."[31]

A lack of empathy is certainly not a universal trait in the world of tech; in recent years, we've seen an increasing number of employees—such as Frances Haugen, Timnit Gebru, and Sophie Zhang—who have come forward with deep concerns about the social impact of their work.

Workers with a more mature outlook may opt out of risk-engendering projects, as we're now seeing in the A.I. field. According to Kelsey Piper, "There are so many people, very good machine learning researchers who I know, who nobly made the decision to not play this game because they thought they might kill us all. And so they are out of this game, and they are working in academia on much smaller models or working on regulatory frameworks."

This kind of responsible behavior is great to witness, but if the people who show it opt out, a field can become dominated by less judicious people. Piper says that the people she mentioned are "doing good work, but they're not the people at these labs, because they said no, and they walked away. And so we have the people who, for whatever reasons, didn't walk away."[32]

Compared to what we saw with earlier digital technologies such as social media, though, there is one bright side to the current situation with A.I. This time around, there has been a much more robust discussion, within and outside the Valley, about possible risks and downsides.

Before we move on from psychological causes, I'd like to mention one more issue. A lack of empathy and discomfort in the presence of other people can be symptoms of autism spectrum disorder, which—in its milder range—can manifest as Aspergers syndrome. Katy Cook has written that, "There is much anecdotal evidence and growing research that points to a correlation between the type of work necessitated in tech and the analytical, highly intelligent, and cognitively-focused minds of 'Aspies' who may be instinctively drawn to the engineering community."[33,j] Colorado State University professor Temple Grandin, perhaps the best-known spokesperson for people on the spectrum (and a person with autism herself), has also made this connection, saying that "half of Silicon Valley has something you'd call Asperger's."[34,k]

On the other hand, Cook points out that "a lack of emotional awareness, dysfunctional patterns of relating, and a lack of empathy" are actually characteristics of *alexithymia*, a disorder that tends to get conflated with autism. (Only about half of those on the autism spectrum have both.)[35]

For whatever reasons, some engineers and developers exhibit a

---

[j] Many people on the spectrum also exhibit positive traits such as honesty, reliability, persistence, and highly developed talents. A big shout-out to Greta Thunberg, not to mention Carl Sagan, Anthony Hopkins, and—possibly—Albert Einstein, W.B. Yeats, Michelangelo, and Mozart. (https://www.onthespectrumfoun dation.org/famous-people-with-asperger-s)

[k] Peter Thiel has also commented on a prevalence of Asperger's in Silicon Valley (as quoted in https://www.newyorker.com/magazine/2011/11/28/no-death-no-taxes), and Elon Musk has publicly said that he has been diagnosed with Asperger's.(https://www.washingtonpost.com/arts-entertainment/2021/05/09/elon-musk-hosts-snl/)

discomfort with the presence of other people, and a number of companies in Silicon Valley have even seen this as an indication of the right mindset for the job. In order to create aptitude tests for potential hires, the computer industry commissioned a couple of psychological research projects back in the 1960s. A Vocational Interest Scale for Computer Programmers[36] suggested two key desirable traits: "an interest in solving puzzles and a dislike of or disinterest in people." This influenced the design of the tests, which Cook says "were used to select engineers within the industry for decades."[37]

A desire for distance from other people can affect what tech creators want to spend their energies developing, and artificial intelligence seems like a prime example. Joi Ito, a former director of the Media Lab at M.I.T., has said that, "One of my concerns is that it's been a predominately male gang of kids, mostly white, who are building the core computer science around A.I., and they're more comfortable talking to computers than to human beings. A lot of them feel that if they could just make that science-fiction, generalized A.I., we wouldn't have to worry about all the messy stuff like politics and society."[38]

Jaron Lanier put this more starkly, when he wrote that, "We technologists are ceaselessly intrigued by rituals in which we attempt to pretend that people are obsolete."[39]

Every time I ride the New York subways, I see ads for tech startups that promise some ultra-convenient new service. Some commentators[40] have remarked that there's something oddly adolescent about these apps, as if they were created by young men trying to recreate the things their mothers did for them: bring them food, wash their clothes, drive them around.[l]

On the other hand, one might speculate that the fact that Silicon Valley workers are often male and childless might have made it harder for many of them to appreciate the impacts that their tech could have on kids—and to care enough about those impacts to design their creations to be safer for them.

In any case, when you put a lot of young people inside a cultural bubble and give them countless job perks and lots of money, it's little wonder that they sometimes design technologies that don't address serious needs.[m] Don Norman, head of the Design Lab at U.C. San Diego, has said that the first thought of too many designers and technologists is "let's make an app." He explained that, "A lot of the companies don't think much of the implications of what they're doing—they're often very excited by the technology, it's all about the technology and what the technology can do. In the design field, we try to train our students to find out what the fundamental need is for people first, then decide how to deal with that and what they should do. The technology comes last."[41]

Of course, in the world of venture capital, the most common question is not, "Does this startup address real human needs?" All too often, it's *Can this app be scaled up quickly and then make us a*

---

[l] This might seem relatively harmless, if not for the fact that so many of these apps seem premised on putting other people out of work. I think of an ad for an online banking startup that read, "2020 [the start of the Covid-19 pandemic] Proved a Lot of Things, Like Bank Branches Aren't Necessary." I wonder if it ever occurred to the founders that bank branches provide jobs for human beings.

[m] Perhaps the most notorious example was the Juicero, made by a startup whose founders raised $120 million of venture capital funding so they could try to get consumers to buy a $400 machine to dispense drinks made from packets of diced fruits and vegetables— until people realized that they could simply squeeze the packets into glasses by hand.

ton of money in several years when the company goes public, or we sell it?

Robert Gardiner cautioned me about emphasizing psychological issues over economic considerations. "I think there's a really big gray area when you look at what's a lack of emotional intelligence, and what's just capitalism. There may be many other factors at work, like cost reduction. In the design process, [you may hear that] 'One way is cheaper and easier to implement, and gives us access to a larger market, so we'll go in that direction.' That tends to be how some of those decisions get made."[42]

Silicon Valley, after all, is a brutally competitive industry with intense commercial pressures. In many cases, what we're seeing may not be emotional immaturity, but rather a myopic focus on practical issues. Engineers race to solve technical problems before competitors at other companies can get there first. Growth and Ads team members fret about meeting quarterly quotas. Founders obsess about securing more venture capital before they burn through their assets. It's not easy to take the time and energy to step back and consider the bigger picture, to anticipate consequences that might go beyond the immediate technical or financial issues.

Commenting further on the lack of empathy, Om Malik argued that, "It is not that all players are bad; it is just not part of the thinking process the way, say, 'minimum viable product' or 'growth hacking' are."[43]

The story of one worker named Wendy Liu, as she tells it in her admirably honest, soul-searching memoir *Abolish Silicon Valley: How to Liberate Technology from Capitalism*,[44] provides a powerful ground-level view of what can happen when a tech developer is short on empathy.

The young Canadian software developer moved to San Francisco in 2013 to do an internship at Google, and then spent several years trying to co-found a tech startup. She and her little band of partners initially planned to develop data analysis software that they could sell to advertising companies. They had a tough time attracting venture capital, though; their concept was vague, and they had to compete with other founders who were creating similar products.

Some people come to the Valley determined to develop one product that they fervently believe in, but Liu and her friends didn't have that; after their initial concept didn't gain traction, they flailed around trying come up with a marketable idea. They considered getting into automation software, first for call centers, then for fast-food restaurants. At first, Liu convinced herself that this could be a socially valuable project. "In my diary," she later recalled, "I jotted down a tentative mission statement: 'transition humanity to a state where human labour is valued and is harnessed to its full potential.' That meant no more wasting human time on repetitive, easily automatable tasks; instead, machines could be used to liberate people from necessity."[45]

As she watched an actual worker at a fast-food restaurant down the street from her office, though, Liu began to wonder about the impact that automation might have. (In other words, some empathy kicked in.) Yet she still managed to rationalize what she was planning. "What would happen to the employee? Well, she could find another job. Of course, other jobs were getting automated or made more unpleasant through the intensification of surveillance technology, but the plight of this employee was not my problem. My only responsibility was to my technology, my investors, the higher notion of progress; the human beings at the short end of the stick were inconveniences to be automated away, necessary sacrifices at the altar of efficiency."[46]

"The technical problem captivated me," she concluded. "I could worry about the ethics later."[47]

She stayed focused on creating software that would make companies more efficient and profitable, but she found more empathy arising when she met a struggling actor who eked out a living as a delivery worker. Encountering people working in such harsh conditions began to trouble Liu more and more, and then she had an epiphany. "It finally hit me what I'd missed the whole time we were thinking about automation: no matter how good our automation technology, there was no viable business model in which we would actually achieve our goal of giving workers back their time. We wouldn't be selling to the *workers*, after all; instead, we would sell our technology to the company, which would then implement it in a way that would allow workers to be fired."[48]

Once she was able to empathize with other, less privileged workers, Liu realized that she didn't want to contribute to creating technology that would devalue them. She began to read critiques of the surveillance capital system, which made her reflect on her first attempt at a startup:

"Why had we thought that it was OK to sell personal data without consent so that brands could target people better? I looked back through my earliest diary entries, and it was almost physically painful to see what I chose to record on a day-to-day basis: the number of users we tracked, the attributes we could infer [...]. The thing is, I had always suspected, deep down, that what we were doing was wrong. Our business model had never felt particularly respectable, nor did it feel like it contributed to society in a way that legitimated even the little revenue we got. [...] The people on whom we counted for advice or money or connections never raised an ethical objection to our flagrant disregard for other people's privacy. Why would they? We had seen a niche in the market, and we tried to fill it. The market was the ultimate arbiter. Questions of ethics were beside the point."[49]

For years, Liu had been fiercely driven. She had given up her social life and spent all of her time writing code—she had the right personality to eventually succeed in the Valley. "In the past," she recalled, "I had believed that it was enough for me to work on an interesting technical problem. It didn't really matter what the technology actually *did*, as long as I was learning something and getting paid."[50] In the end, though, she was so troubled by the amoral nature of this culture that she dropped out of it, went to graduate school to study the social science of inequality, and became an activist for tech workers.

In the last few years, those workers have faced a harsh, upsetting wakeup call. Before the recent *techlash*, the public looked on their industry as a remarkable fountain of free apps and services. It must have been quite dispiriting to hear a steady drumbeat of whistle-blower testimony and other complaints about their companies. and to watch public opinion take a dramatic disapproving turn. Although undoubtedly the vast majority of them never had any intention to cause harm, they could no longer avoid the fact that—while they were focused on tackling engineering challenges or meeting quotas—some of their creations had caused major problems for other people.

As Cook told me, "People didn't get into tech thinking that it would be something so socially important that they had to worry about these kinds of issues. Like if you're really good at network security, then you could go to work for Tesla. And [then] there's a car crash that kills someone with automatic driving. All of a sudden, you're attached to this human issue that as a coder, you probably never thought you would have to deal with."[51]

Thankfully, the shock of recent revelations has prompted many workers to take a stronger ethical stance. And tech executive and consultant Maëlle Gavet, author of a book called *Trampled by Unicorns: Big Tech's Empathy Problem and How to Fix It*, says that some companies, such as Microsoft,[52] don't just use empathy as a buzzword, but take concrete steps to actively promote it within their workforces; she says this commitment is manifest in such areas as the operations of their Human Resources department, including hiring and promotion.

For some corporate leaders, however, the word connotes an emotional mushiness; they may believe that it gets in the way of making "hard-headed" business decisions. Gavet offers a useful distinction:

"I think the people who make that argument tend to confuse 'affective empathy' with the cognitive variety I'm advocating. Affective empathy implies sharing, almost physically, the feelings of other people. It makes it harder to share direct feedback (you don't want to hurt people's feelings) and make tough calls (you want everybody to feel good about a decision). Cognitive empathy, on the other hand, helps you understand how other people feel and think, and as a result you adapt your decisions and behavior accordingly; it thus enables better-informed decision-making. As a leader, you should still act in the best interest of your business, but by understanding how your decisions affect other people, both positively and negatively, you're better able to act with clarity and decisiveness, with fewer negative side effects."[53]

Unfortunately, understanding how his decisions affect other people is not one particular tech mogul's strong suit.

For decades, Silicon Valley moguls shied away from taking a direct role in politics, preferring to participate mainly through lobbying for deregulation of their own industry, and donations to candidates who might take a hands-off approach. That changed radically with the reelection of Donald Trump in 2024, which paved the way for an astonishingly abrupt and forceful rampage by Elon Musk and his team of marauding DOGE engineers (young, white, male) through the ranks of U.S. government, as he indiscriminately fired tens of thousands of people who had devoted their lives to public service, from forest rangers to weather forecasters, from cancer researchers to scientists developing prosthetics for veterans.[54]

It should have been easy to imagine the pain of people who had been capable and helpful civil servants for years, only to find that they had been deprived of their livelihoods without any warning and for no good reason, and suddenly had to panic about how they were going to put food on their family tables, yet Musk wielded his (literal) chainsaw with oblivious glee.[55] He also provided the extremely unsavory spectacle of the world's richest man cutting off aid to sick and starving people around the world.

To be a worthy leader in a democracy, you have to be able to envisage the effects of your decisions on the population, and strive for the common good. In 2025, what could not have been clearer was that Elon Musk could not have cared less about those he was causing to suffer, giving the lie to his smirking claim that he was "pretty good at running human in emulation mode."[56]

Chapter 7

On History, And Progress

"There was a strong sense back then—certainly you heard it from Mark and the people around him—that wiring the world was good in and of itself. There was a widespread belief in the inevitable forward march of history. I don't know that that came from books, or from anywhere in particular—I think it was just understood."[1]

—Chris Hughes, Mark Zuckerberg's college roommate and one of Facebook's first employees

As we'll see in this chapter and the following three, the Wheel is profoundly shaped not just by the culture and psychology of tech creators, but how they think about technology. That mindset is largely shaped by their view of the past.

Earlier, I offered a quote from an insider who ruefully pointed to a lack of a sense of history in Silicon Valley. Here's one from an insider who has actively argued that such knowledge is unnecessary. Anthony Levandowski, a prominent former engineer at

Google and Uber, told an interviewer that, "I don't even know why we study history. It's entertaining, I guess—the dinosaurs and the Neanderthals and the Industrial Revolution, and stuff like that. But what already happened doesn't really matter. You don't need to know that history to build on what they made. In technology, all that matters is tomorrow."[2]

On the other hand, many tech leaders *do* have a historical perspective. Unfortunately, it's often a distorted one.

According to this version, during our first 300,000 years we *Homo Sapiens* lived brief, harsh lives as hunter-gatherers. The discovery of fire made our lives a little easier, but we still had to wander the countryside and spend all our time in a desperate struggle to find each day's food, and fight off wild animals with pointed sticks.

Some 11,500 years ago, after a long Ice Age, the earth finally began to thaw. Around that time, the smartest primitives noticed that edible plants seem to grow where seeds fell on the ground. They thought *Hey, wouldn't it be great if we could stop chasing these animals around and get our food from this easier, more predictable source? What if we take some of our pointed sticks and use them to dig up the soil, then place some seeds down in it?*

When these nomads made the revolutionary switch to becoming farmers, using the technologies of agriculture allowed them to settle in fixed places and build up storable surpluses of food, which meant that not everyone had to devote their lives to finding it, which allowed some people to specialize in other areas—to become priests and soldiers and artisans and kings. A need for complex irrigation systems led to political hierarchies that could oversee their construction, and to centralized bureaucracies, whose functionaries invented writing and mathematics as a way to keep track of the extra food.

Rovers could only carry the barest of essentials, but settled humans were able to gather possessions, develop more complex tools, and invent architecture. Whereas each square mile of land was only able to support a few hunter-gatherers, agriculture and animal husbandry allowed populations to grow dense, which led to the rise of cities, and the creation of legal and political systems, and powerful states. The societies that developed the most advanced technologies, such as ancient Egypt and Rome, were able to conquer more primitive cultures and build grand monuments such as the pyramids of Giza and the Colosseum.

After a pause of some centuries during the Dark Ages, when the world was overrun by barbarians and nothing much happened, we had a re-flowering of technology and culture, mostly in Europe, with the Renaissance and the Enlightenment, and were able to enjoy such glories as the Sistine Chapel and Beethoven's symphonies. The Industrial Revolution,[a] centered in Britain, brought a boom in technologies that radically improved our standard of living, and then the United States became the center of innovation, and then we got to our current world, full of high-tech miracles.

In his seminal book *Against the Grain: A Deep History of the Earliest States*, political scientist and historian James C. Scott summed up this traditional view: "No one, once shown the techniques of agriculture, would dream of remaining a nomad or forager. Each step is presumed to represent an epoch-making leap in mankind's well-being: more leisure, better nutrition, longer life expectancy, and, at long last, a settled life that promoted the household arts and the development of civilization."[3] To sum up

---

[a] There were actually several of these. The first lasted from the mid-1700s to about 1830 and took place largely in Britain, which benefited from sitting on top of huge coal deposits that provided fuel for new machines, such as steam engines. The second, more global, lasted from around 1870 to 1914, and featured the use of gas and electricity, and mass production. The third featured the invention of electronics and computers, nuclear energy, and the internet. (Some historians break these periods down into even smaller increments.)

this story, the technologies of agriculture initiated a grand Ascent of Man, which would eventually become centered in Europe and then the U.S., and continues inexorably today.

Jason Crawford, a former software engineering manager and tech startup founder, felt so inspired to promote this view that he started a blog—and a nonprofit organization—called *Roots of Progress*. Here's a sample of his writing on the subject:

> "50,000 years ago our ancestors lived at the mercy of nature." [...] Their lives were characterized by abject poverty, superstition born of ignorance, and constant tribal warfare. We have come a very long way. We live in buildings, not caves. We are surrounded by mass-manufactured products made of steel, glass and plastic. We extract vast sums of energy buried in the ground and we make it do our bidding. We have all the food we could want—so abundant and delicious that we have to restrain ourselves from eating too much. We have antibiotics and laser surgery. We zip around the world at hundreds (and maybe soon thousands) of miles an hour. We can communicate with anyone, anywhere, anytime, instantly. And all of this is made possible by a vast and rapidly expanding scientific knowledge of the world. Further, we live in relative peace and freedom, made possible by governments that maintain law & order—and the institutions of democracy and republicanism that keep those governments in check. In short, compared to our prehistoric ancestors, we are *smart, rich and free.*"[4]

That's pretty much the account of human history that I learned in elementary school—perhaps you did too. And it seems to be the version that informs the techno-optimistic worldview of many of today's tech titans.

～

That's certainly an impressive tale, but—unfortunately for its proponents—a number of archaeologists, anthropologists, and historians have found that a good deal of it is flat-out wrong.

First, it's skewed in a number of significant ways, including the fact that it's too Western-centric. As Lewis Mumford pointed out, "The windmill perhaps came from Persia in the eighth century. Paper, the magnetic needle, gunpowder, came from China, the first two by way of the Arabs; algebra came from India through the Arabs, and chemistry and physiology came via the Arabs, too, while geometry and mechanics had their origins in pre-Christian Greece."[5]

It's also easy to think that technological innovation didn't really get cooking until that first Industrial Revolution, but Mumford took pains to point to a great variety of technologies that inventors around the world had already developed, including water wheels, clocks, telescopes, magnetic compasses, metallurgy, mining, sailing technologies, and the scientific method itself. He was moved to conclude that, "So far from being unprepared for in human history, the modern machine age cannot be understood except in terms of a very long and diverse preparation. The notion that a handful of British inventors suddenly made the wheels hum in the eighteenth century is too crude even to dish up as a fairy tale to children."[6]

In *The Dawn of Everything: A New History of Humanity*, a seminal 2021 book by David Graeber, an American anthropologist, and David Wengrow, a British archaeologist, they also take exception to the traditional Western view of history. They note that, "In the Middle Ages most people in other parts of the world who actually knew anything about northern Europe at all considered it an obscure and uninviting backwater full of religious fanatics who, aside from occasional attacks on their neighbors ('the Crusades'), were largely irrelevant to global trade and world politics."[7] What they *could* soon boast of was better military technology, which enabled them to colonize a number of other global empires.

Historians have also tended to focus on male inventors, and to think of technology as a male-centric enterprise. Mumford noted that we "identify tools and machines with technology," ignoring a host of important invented containers, such as "hearths, pits, houses, pots, sacks, clothes, traps, bins, byres, baskets, ditches, reservoirs, canals, cities."[8] Historian Carroll Pursell argues that this may reflect a masculine bias.[9]

Graeber and Wengrow warn of the same slant: "Instead of some male genius realizing his solitary vision, innovation in Neolithic societies was based on a collective body of knowledge accumulated over centuries, largely by women, in an endless series of apparently humble but in fact enormously significant discoveries. Many of those Neolithic discoveries had the cumulative effect of reshaping everyday life as profoundly as the automatic loom or lightbulb."[10,b]

We tend to see human history and the history of technology as one and the same, a series of eras when people used ever stronger materials to build ever more powerful tools and machines. The Iron Age, the Industrial Revolution, the Digital Era. For obvious reasons, archaeologists have focused on the cultures that left behind stone monuments and metal tools and weapons, and discounted those whose achievements left less physical evidence. Too often, our view of the past resembles a selective x-ray of the human body which registers only metal implants, leaving out the blood and sinews and heart.

History, as the saying goes, is written by the winners. It was also written by those who had developed the technologies of writing and printing, omitting highly developed early cultures in

---

[b] I need hardly point out that our society continues to devalue women's contributions to technology—see Silicon Valley today.

places like Australia, the Pacific Islands, subequatorial Africa, and most of the Americas.[11]

We often think of Real History as beginning with the use of stone tools and the dawn of agriculture, leading to what Graeber and Wengrow call "an odd insistence that for many tens of thousands of years, nothing happened."[12] But then, of course, something very obvious *did* happen: humans invented technologies for growing plants and raising animals. Jared Diamond, the historian who wrote the bestselling book *Guns, Germs, and Steel*, once wrote an essay about agriculture titled "The Worst Mistake in the History of the Human Race."[13]

How on earth could he reach such a contrary conclusion?

First, let's consider those supposedly miserable hunter-gatherers. Recent studies have shown that they spent less time[14] and energy[15] procuring food than farmers and herders, and had more hours to relax, hang out with each other, and enjoy nature. And did agriculture provide a more stable source of food? Not necessarily. Hunter-gatherers collected a great variety of foodstuffs from many sources, whereas, in the event of drought or blight, it could be fatal for farmers to rely on just a few crops.[16] Hunter-gathers also ate a healthier, more balanced, and more abundant diet, suffered less malnutrition, and were "several inches taller, on average, than early farmers."[17] What's more, their lives featured far less drudgery, and were experientially richer and more varied.[18]

Crawford claims that they lived in "abject poverty" and "constant tribal warfare." Actually, if they had enough food and were reasonably healthy, it's unclear how they can be called poor. (Compared to whom?) And the impression that they were constantly fighting seems to be misleading, likely founded on the work of a few early anthropologists who happened to study particularly warlike groups such as the Yanomamö people of South

America. There is currently some dispute about the issue,[c] but archaeologists have found little evidence of what could be called war before the advent of agriculture.[19] Anthropologist Douglas P. Fry and developmental psychologist Patrik Söderberg concluded that, "The nomadic forager default is getting along with neighbors, not warring with them. Of any lessons we derive from nomadic forager studies about war, peace, and human nature, this one is the centerpiece."[20]

The spread of agriculture and animal husbandry *did* enable greater density of populations—which brought its own new problems, often health-related. Diamond notes that, "The major killers of humanity throughout our recent history—smallpox, flu, tuberculosis, malaria, plague, measles, and cholera—are infectious diseases that evolved from diseases of animals."[21]

John J. Ratey, a doctor and professor at Harvard Medical School, and journalist Richard Manning wrote a book called *Go Wild: Eat Fat, Run Free, Be Social, and Follow Evolution's Other Rules for Total Health and Well-Being.* They describe what have been termed "diseases of civilization," which often came with the new sedentary lifestyle and its great reliance on dietary starch (which turns to glucose).[22] These health problems, such as obesity, high blood pressure, and strokes, were largely absent in hunter-gatherers. They also cite "diseases of longevity," noting that "the blessings of Western civilization, especially controlling communicable diseases, made people live longer and so gave them more time to develop heart disease, cancer, and type 2 diabetes."[23]

The first humans to domesticate animals eventually developed immunity to some of their diseases, yet they carried those diseases to other cultures, decimating the populations of early Native

---

[c] It's quite telling of our current state of extreme partisanship that a debate about this now rages not just between anthropologists, but between today's political hawks and doves, in attempts to justify their own ideologies.

America, Australia, South Africa, and the Pacific Islands.[24] And farmers sometimes caused problems that could collapse their *own* societies, such as deforestation leading to siltation of waterways and flooding, and to severe salinization of the soil.[25]

The hunter-gathers are often described as egalitarian. They didn't "own" land and couldn't accumulate property, so no one could grow rich, or consolidate great power. But—in this revised view of history—agriculture enabled surpluses, and soon we had authoritarian leaders like chieftains and kings who were able to impose taxes on that bounty, and become wealthy, and develop armies and warfare to steal riches from other societies. What's more, they were able to force others to produce that surplus through many forms of forced labor, ranging from corveé[d] to serfdom to outright slavery.[26] As Scott observes, "It would be almost impossible to exaggerate the centrality of bondage, in one form or another, in the development of the state until very recently."[27,e]

In sum, a technology invented to make it possible to grow food had the unintended consequences of the privatization of property, social and economic inequality, disease and ill health, slavery, and war.

These versions of history are often framed as a sort of intellectual sparring match between two 17th century philosophers. The Englishman Thomas Hobbes wrote that human life in an original State of Nature must have been "solitary, poor, nasty, brutish, and

---

[d] Corveé is intermittent, unpaid labor exacted by rulers, often for the purpose of public works projects like building pyramids or roads.

[e] He cites a staggering statistic that "as late as 1800 roughly three-quarters of the world's population could be said to be living in bondage." (*Against the Grain*, pp. 155-156)

short,"[28] but that it was then substantially improved through the development of civilization. The Swiss Jean-Jacques Rousseau propagated a vision of early Noble Savages living in a state of Edenic simplicity and bliss, which was ruined by the decadent, corrupting influence of civilization. In other words, the former saw history as a story of human life getting dramatically better, and the latter as a tale of it going dramatically downhill.

Why have I spent so much time here talking about the long-ago past? Because the first version fuels the rampant techno-optimism that has led so many tech moguls to ignore possible downsides to what they're doing. **Wheel factor: those who paint an overly rosy picture of the past are likely to also paint an overly rosy one of the future.**

On the other hand, the second version, if taken too far, can lead to a generalized technophobia—as exemplified, for example, in the dark 35,000-page manifesto of one Theodore John Kaczynski, a.k.a. The Unabomber.

As we'll see in later chapters, social scientists such as Graeber, Wengrow, and Scott have recently offered a third, surprising version of history, considerably more interesting and complex. It may suggest a better route going forward in terms of how we relate to our new technologies.

For our purposes now, though, let's delve a bit deeper into the first version of history, since it seems so influential among today's tech movers and shakers. As political theorist Langdon Winner summed it up: "The orthodox view stresses a few key notions: that the conquest and control of nature is modern society's great mission; that efficiency is a universally applicable criterion of social choice; that modern history shows a linear accumulation of scientific progress in human well-being; that our individual and collective mission is a never-ending quest to remain competitive, racing forward along technology's cutting edge. In today's parlance, the glossy term to describe commitments and projects of this kind is 'innovation'."[29]

At the center of this worldview lies the notion that advances in technology drive something called Progress.[f]

Jason Crawford likes to cite Harvard psychologist Steven Pinker, perhaps the best-known current proponent of the idea that technological Progress has brought us to the pinnacle of human history. Both men are fond of graphs showing how things like World Gross Domestic Product and Average Life Expectancy have risen in the modern era.[g] Indeed, it would be churlish to ignore the fact that humanity's overall situation has improved in terms of a number of metrics, including infant mortality, access to education, or percentages of people living in bondage or extreme poverty.[h]

Listening to such cheerleaders, we might easily suppose that the desire for Progress has always been a fundamental driver of human civilization. In reality, however, it's just a concept and—

---

[f]   To emphasize its mythic status, I'm giving it a capital "P."

[g]   Some argue that these graphs have been selectively chosen to paint an overly rosy picture. See, for example, "Steven Pinker's Ideas Are Fatally Flawed. These Eight Graphs Show Why" by Jeremy Lent https://www.opendemocracy.net/en/transformation/steven-pinker-s-ideas-are-fatally-flawed-these-eight-graphs-show-why/)

[h]   In his book *Small Is Beautiful: Economics as if People Mattered*, E.F. Schumacher wrote, however, that the modern Western economist "is used to measuring the 'standard of living' by the amount of annual consumption, assuming all the time that a man who consumes more is 'better off' than a man who consumes less. A Buddhist economist would consider this approach excessively irrational: since consumption is merely a means to human well-being, the aim should be to obtain the maximum of well-being with the minimum of consumption." (p. 53)

like "privacy" or "childhood"[i]—it was a notion that had to be invented.

That may seem hard to believe. Didn't people always see life as a struggle for improvement, resulting in a rising line of general betterment, especially through technological means?

Not in many early cultures. As Graeber and Wengrow point out, "In traditional societies, according to [historian of religion Mircea] Eliade, everything important has already happened. [...] People living in this mental world, he felt, saw their own actions as simply repeating the creative gestures of gods and ancestors in less powerful ways, or invoking primordial powers through ritual."[30] The Christian tradition has long seen human existence as a struggle to return to an earlier, idyllic Eden. Other cultures, such as the Hindus, pre-Christian Greeks, Chinese, and Aztecs, saw life as a repeating cycle, in which we rise up, do our thing for a while, and then die out, only to start all over again.[31]

According to Lewis Mumford (writing in 1934), "Not only is this notion of progress a European—and, by extension, a North American—phenomenon, it is also a relatively recent one. Physicist and novelist Alan Lightman reckons it is less than 300 years old, and connects it with the Industrial Revolution."[32]

Of course, if you want to claim that technology has borne human beings on a steadily more positive path, it's hard to deny that the chariot has hit significant potholes.

If Progress was consistent, after the Enlightenment we should have become even more sane and, well, *enlightened*, yet the 20th

---

[i] Some argue that the modern Western conception of childhood dates only to the 18th century and the Enlightenment. (See *Centuries of Childhood* by historian Philippe Aries, and *The Invention of Childhood* by historian Hugh Cunningham.)

century was marked by frenzies of violence on an astonishing, unprecedented, worldwide scale. Humans invented—and eagerly deployed—tanks, machine guns, land mines, nerve gas, gas chambers, napalm, and nuclear weapons, and killed over two hundred million of their fellows through wars, other conflicts, and genocides.[33] Like a theologian struggling to avoid discussing the Holocaust, Crawford calls the World Wars "outliers,"[34] ignoring the fact that the history of technology has always been, in significant part, a story of humans getting better at killing each other.[j]

The story of technology has taken other dark downward turns. The first Industrial Revolution certainly brought great technical advancements, but it also featured horrendous pollution and a great deal of miserable adult and child labor, deep down in mines or in what William Blake described as "dark Satanic mills." The latter supplied a vast new market for machine-produced clothing. Steven Johnson wrote that, "Between slavery and the grotesque working conditions of early industrialization, one can make the argument that the desire for cotton was the single worst thing to happen to the planet between 1700 and 1900—that it provoked more suffering than any other new development in that period. Those early shoppers savoring the soft fabrics and vibrant patterns of calico at the end of the seventeenth century would have been baffled to know that their fashion decisions were going to unleash such traumatic and deadly forces across the globe."[35]

Wheel factor: the assumption that the develop-

---

[j]   To be fair, military organizations have also funded a plethora of inventions with nonlethal, very useful applications, such as synthetic rubber, the jet engine, radar, satellites, microwave ovens, digital photography, GPS, and—of course—the internet, not to mention duct tape, T-shirts, sanitary napkins, and super glue. https://interestingengineering.com/innovation/9-military-spin-off-technologies-we-use-almost-everyday;https://www.usatoday.com/story/money/2019/05/16/15-commercial-products-invented-by-the-military/39465501/

ment of technologies necessarily means progress for humanity.

That era marked the apotheosis of what happens when you combine the unbridled development of technology with an unregulated economic system.[k] Many techno-optimists prefer to ignore that aspect of its history. As author Wayne Grady puts it, "Since the Industrial Revolution, postagrarian societies have no longer looked back to the past for inspiration, but forward to the future. [...] It is our love affair with progress, and our equating of progress with technology that has made the past seem irrelevant; the past had less (or slower) technology, we think, and was therefore primitive."[36]

Primitive; when past historians mentioned cultures with low levels of technology, they also tended to speak of "barbarians" and "savages." We're more sensitive about such terms these days, and may no longer speak about technologically superior conquerors from Europe bringing Civilization to the benighted residents of other cultures—yet we haven't totally abandoned the notion. As Graeber and Wengrow point out, even when Jean-Jacques Rousseau was lauding his blissful "Noble Savages," he still made the mistake of thinking that they were mentally simple.[37]

It's interesting to note that when French explorers visited the eastern shore of North America in the 17th and 18th centuries, they were struck by the oratorical and intellectual abilities of the natives. Graeber and Wengrow quote a Jesuit priest who said of the Wendat tribe, "as regards intelligence, they are in no wise inferior to Europeans," and they note that, "Some Jesuits went further, remarking—not without a trace of frustration—that New World

---

[k] It took decades before governments began to catch up and start doing something to rein in the extremely long labor shifts, dangerous working conditions, and devastation of the water and the air.

savages seemed rather cleverer overall than the people they were used to dealing with at home."[38]

In one of the most fascinating sections of *The Dawn of Everything*, they show that during the Colonial era, Native American tribes lived in societies that were far more politically egalitarian and free than those of their visitors from across the Atlantic. "In the considered opinion of the Montaignais-Naskapi," they write of one Canadian tribe, "the French were little better than slaves, living in constant terror of their superiors."[39] They quote a French observer of the time, who noted of the Wendat tribe that, "They reciprocate hospitality and give such assistance to one another that the necessities of all are provided for without there being any indigent beggar in their towns and villages."[40]

The French brought some of the Native Americans home so they could witness European civilization in all its glory. Instead, they were shocked by the harsh, punitive judicial systems, the physical uncleanliness of the people and their cities, the citizens begging in the streets, and the ways in which wealth seemed to arbitrarily confer the power to coerce and abuse others.[41]

Ultimately, Graeber and Wengrow argue that some of the roots of democratic thought that we credit to Europeans of the Enlightenment might have passed from the natives of North America's eastern shore back to Europe, rather than the other way around.[42]

In any case, we tend to equate greater technology with greater intelligence. We draw snide cartoons of early humans scratching themselves and grunting, yet they lived in a profound connection with their natural environment that we have largely lost today. Without access to writing or computers, they had prodigious memories, and immense knowledge of plants and animals and weather and terrain. Ratey and Manning speak of "the hunter-gatherer state of mind, a hyperawareness, a presence, a capacity for observation we can only begin to imagine."[43]

I often think of that quote as I walk to and from work, passing legions of modern humans stumbling along the sidewalks as they stare down at their phones.

In contrast to the desolate existence imagined by Crawford, our ancestors often enjoyed good health and strong human bonds, and they lived without rulers ordering them around or stealing their stuff. What if—technology-deprived as they were—they were as happy as we are today?

What if they were *happier*?

Chapter 8
How We Think Of Tech

Proclaiming what's going to happen is a popular way to shrug off taking responsibility for helping to determine what's going to happen.

—Rebecca Solnit[1]

THE IDEA THAT THE DEVELOPMENT OF TECHNOLOGY NE-cessarily equals progress for humanity is just one of a number of questionable notions that permeate today's world of tech.

As I mentioned earlier, one of the things that inspired me to write this book was Neil Postman's quote about how once we build machines, they get minds of their own. I've come to recognize one big flaw with this quote: no past inventions have been capable of behaving as independent agents. There's a term for the false notion that they can: ***technological autonomy***.

It's one thing to encounter this fallacy in a few student papers; it's another to see it pervading the pages of Kevin Kelly, one of the most prominent authors of books about tech.

Here's a sample from one of them: "Looking back, I think the computer age did not really start until this moment, when computers merged with the telephone. [...] All the enduring consequences of computation did not start until the early 1980s, that moment when computers married phones and melded into a robust hybrid. In the three decades since then, this technological convergence between communication and computation has spread, sped up, blossomed, and evolved."[2]

Look at all of that merging and blossoming and marrying and evolving! What's missing here? In Kelly's account, there's a complete absence of human beings—it's as if all this just happened without us.

The author has even published a book accompanied by this cover tag line: "Technology is a living force that can expand our individual potential—if we listen to what it wants."[3] (If you want to test the logic here, imagine if we had a legal system in which we didn't identify criminals, but talked about crimes being committed because *that's what crimes want*.)

Lewis Mumford recognized this fundamental error when he wrote—almost a century ago—that, "Technics and civilization as a whole are the result of human choices and aptitudes and strivings, deliberate as well as unconscious, often irrational when apparently they are most objective and scientific: but even when they are uncontrollable they are not external."[4]

We speak of Frankenstein, but it's essential to remember that he was the inventor, not the creature that he built. *We* make technology; it doesn't make itself—at least, not until artificial intelligence models begin moving beyond our control.

Timnit Gebru warns that speculation about that possibility can be distracting because it "ascribes agency to a tool rather than the humans building the tool. That means you can abdicate responsibility: 'It's not me that's the problem. It's the tool. It's super-powerful. We don't know what it's going to do.' Well, no—

it's *you* that's the problem. You're building something with certain characteristics for your profit."[5]

Big Tech leaders commonly blame the tool instead of its creator as a way to avoid responsibility, but they also often hide behind an opposite argument. They love to claim that technologies are neutral, neither good or bad—that these moral values only arise from the ways in which people use them.

This is a favorite argument of gun manufacturers. It's also a blatant fallacy. A neutral technology is a rare thing, because so many inventions carry within them potential for harm. (This is obvious with an atomic bomb or a malware script, but it's just as true for other technologies—such as social media platforms or A.I. chatbots.)

As I mentioned earlier, technologies never spontaneously emerge into the world, like forces of nature inevitably blossoming into existence. (Advanced A.I. models, for example, can only come into being through the concerted effort of large numbers of human workers, and the conscious application of huge amounts of money and effort.) When technologists release such creations into the world, they enable hazardous potentials that would not be otherwise possible.

As poet William Butler Yeats once put it, "In dreams begins responsibility."[6]

If a creator is conscientious about trying to avoid enabling harmful consequences, they'll be more likely to create a beneficial product, but if they ignore that potential, or claim that they're not accountable for it, bad things are much more likely to happen.

On the other hand, those potentials are often just *possible* outcomes.

Another problematic notion is that new inventions inevitably cause certain *actual* results. (E.g., "Agriculture led to the rise of city states.") This is known as ***technological determinism***.

British semiotician Daniel Chandler noted that it "can usually be easily spotted in frequent references to the 'impact' of technological 'revolutions' which 'led to' or 'brought about,' 'inevitable,' 'far-reaching,' 'effects,' or 'consequences,' or assertions about what 'will be' happening 'sooner than we think' 'whether we like it or not'. This sort of language gives such writing an animated, visionary, prophetic tone which many people find inspiring and convincing."[7]

The problem is that if things go wrong, this way of thinking also puts the blame on the technologies, rather than the people who created them. As an example, I would point out that there's a huge difference between saying "Social media led to greater political polarization" and "Engineers created algorithms to increase user engagement, which siloed users into political bubbles and pushed them towards extreme content."

Unfortunately, Kelly and like-minded thinkers often jump from the notion that particular technologies lead us in certain unalterable directions to an even more insidious one: that certain technologies are *themselves* inexorable. This gets the moniker ***technological inevitabilism***.

It would be impossible to overstate the extent of Kelly's attraction to it. "Massive tracking and total surveillance is here to stay," he tells us. "Ownership is shifting away. Virtual reality is becoming real. We can't stop artificial intelligences and robots from improving, creating new businesses, and taking our current jobs."[8]

Imagine that: it's A.I. and robots that are "improving" and will act to take our jobs, not the human beings who are creating them! What's more, humans are evidently helpless to resist the march of tech. In 2016, Kelly even published a book titled *The Inevitable: Understanding the 12 Technological Forces That Will Shape Our Future.*[a]

Recently I had some of my Gen-Z students do some research on A.I. programs that can generate images based on verbal prompts. One thing that surprised me greatly was that—knowing that this generation is extremely tech savvy—I would have predicted that they'd be enthusiastic about this new technology, yet most of them looked on it with suspicion and distaste. Indeed, they seemed to look on A.I. in general that way. I asked why, and they cited how their generation had gone through such bad experiences with social media, plus they were very concerned by how A.I. uses inordinate amounts of energy and water, and stresses the environment.

We looked at some of the problems with the image-generation apps, including the way they replicate biases in the databases they were trained on. (For example, we entered the prompt "thug" into one model, and watched it generate images of black men.)

I had the class conduct a debate, and they came up with strong arguments both for and against the new technology. I was heartened by their engagement with the topic, but one thing *disheartened* me: every student told me they felt that the advancement of artificial intelligence was inevitable. It didn't occur to any of them that they could actively resist it.

Until the techlash of the past several years, it has been hard for anyone to argue against the *You Can't Stop Progress* sloganeers. As professor of science Langdon Winner put it, "Any suggestion that the forward flow of technological innovation be in any way limited

---

[a] I don't have anything personal against Kelly. It's just that he's so fond of spouting such ideas, and that this kind of thinking is so widespread in the world of tech. Shoshana Zuboff calls Silicon Valley "the *axis mundi* of inevitabilism." (*The Age of Surveillance Capitalism*, p. 222)

by an idea of rational or humane planning is certain to evoke a harsh response. Such proposals violate a fundamental taboo."[9]

Unfortunately, even though inevitabilism is based on questionable premises, it often becomes a self-fulfilling prophecy. Once a technology is set in motion, words of warning can seldom derail it or even slow its momentum. Ethicist Wendell Wallach notes that often only a major accident, disaster, or tragedy can alter the trajectory.[10] (Think of the nuclear meltdown at Three-Mile Island and its impact on the adoption of nuclear energy, or the way in which the explosions of the shuttles *Challenger* and *Columbia* slowed space exploration.)

Some technophiles take the notion of inevitabilism yet another step further, into the often-bizarre realm of the ***technological imperative.*** This is the notion that once we can think of a technology, we *ought* to develop it.

Jaron Lanier points to a notably creepy example. Back in 2006, after it occurred to researchers from the University of Massachusetts and the University of Washington that it might be possible to kill an elderly person by using cell phone technology to hack into their pacemaker, they spent the next two years figuring out how to do it. This was an idea that might have never occurred to anyone else, but they apparently bought into the idea of inevitabilism, and thought they were doing the world a favor by warning how it could be done.[11]

Though Kelsey Piper has pointed to the many risks posed by A.I., she notes that, "You can't count on people, unfortunately, to just say this is ludicrously dangerous, we're not going to do it, because I guess, some share of people are just like, this is ludicrously dangerous, and I'm *going* to do it. I guess this is like the classic tale of hubris, in some ways. Like, the oldest known flaw in human nature is you see the tree of knowledge and you're like, yeah, I'm going for it."[12]

Part of that drive comes from an inability to resist a technical challenge. J. Robert Oppenheimer famously said that, "When you see something that is technically sweet, you go ahead and do it and you argue about what to do about it only after you have had your technical success. That is the way it was with the atomic bomb."[13]

The argument for the bomb, of course, was that the U.S. had to rush to develop it, lest the Nazis get there first.[b] Today, those who make this argument have replaced the A-bomb with A.I., and Germany with China. But as Ezra Klein has pointed out, China has proved considerably more willing to restrict the uses of A.I. than we have in the West. He also notes that, "For all the talk of an A.I. race with China, the easiest way for China—or any country for that matter, or even any hacker collective—to catch up on A.I. is to simply steal the work being done here."[14]

Others have posited that if China (or another country) wants to win out over the West, it doesn't need to develop advanced A.I. at all—it can simply watch *us* develop it, and destroy our society all by ourselves.[15]

Here are two last problematic beliefs.

In his 2013 book *To Save Everything, Click Here*, writer Evgeny Morozov disparagingly coined the term ***techno-solutionism***, the notion that most every social problem can be fixed

---

[b] As it turned out, Hitler's regime never actually came close to developing a useable bomb, despite what the leaders of the Manhattan project believed until 1944. After learning this, though, the U.S. military still went ahead and dropped the bombs on Hiroshima and Nagasaki. (https://www.osti.gov/opennet/manhat tan-project-history/Events/1942-1945/rivals.htm#:~:text=As%20it%20turns% 20out%2C%20the,the%20end%20of%20the%20war.)

through the application of technology. Physicist Ursula Franklin also pointed to it when she wrote that the tech mindset "seems to involve an inherent trust in machines and devices [...] and a basic apprehension of people." She observed that humans "are seen as sources of problems, while technology is seen as a source of solutions."[16]

This is a strange and powerful factor within the Wheel: a tendency to put more faith in technology than in human beings.

One of the biggest problems with techno-solutionism is that it can draw energy and resources away from other, more human-centered ways of dealing with problems. For example, researchers have pointed to a "loneliness epidemic," a new global public health crisis that has been affected by a great rise in the number of people living alone.[17] A.I. proponents tout artificial companions as a solution,[18] but if we devote our efforts to creating those, we might not do enough of what others have suggested, which is to bolster social infrastructure in the non-digital world, in areas such as schools and colleges, senior centers, and workplaces.[19] In general, if we put all our faith in technology to solve our problems, then we're *not* putting it into collective human action, such as social services, and education, and government.

Promoters of techno-solutionism even apply it to damage created by tech itself. It seems that Kevin Kelly missed the irony of his argument when he proclaimed that, "I think we can solve the problems technology produces with more and better technology."[20,c]

---

[c] On the other hand, as technologies such as the exploitation of fossil fuels and factory farming cause global warming and endanger our planet, tech titan Elon Musk argues that the way to ensure the survival of humanity is to use more technology to *escape* the problem (rather than solve it), by sending a few select representatives of our species rocketing off to Mars. (Where, of course, we'll be reliant on technology on a level we have never experienced before.)

Techno-solutionism is closely intertwined with one last type of flawed thinking that has already raised its head a number of times in this book. **When it comes to our factors that tend to generate unintended risks, this one is particularly dangerous: *extreme techno-optimism,* which causes technologists to envision only positive consequences for their creations.**

If you'd like a perfect example, just check out this paragraph from "Why A.I. Will Save the World," a breathless ode to the technology penned by Marc Andreessen, one of today's most powerful venture capitalists:[d]

"Think of what it would mean for literally all existing human labor to be replaced by machines. It would mean a takeoff rate of economic productivity growth that would be absolutely stratospheric, far beyond any historical precedent. Prices of existing goods and services would drop across the board to virtually zero. Consumer welfare would skyrocket. Consumer spending power would skyrocket. New demand in the economy would explode. Entrepreneurs would create dizzying arrays of new industries, products, and services, and employ as many people *and* AI as they could as fast as possible to meet all the new demand. Suppose AI once again replaces *that* labor? The cycle would repeat, driving consumer welfare, economic growth, and job and wage growth even higher. It would be a straight spiral up to a material utopia that neither Adam Smith or Karl Marx ever dared dream of."[21]

---

[d] He originally gained fame as the co-author of the Mosaic browser, and co-founder of Netscape. He also gained a lot of attention in 2023 for writing "The Techno-Optimist Manifesto."

On a more modest note, Andreesen claims that "technology introduced into an industry generally not only increases the number of jobs in the industry but also raises wages," ignoring the fact that companies have already used technologies around the world to force workers out of full-time jobs and into the lower-paying gig economy. He discounts other evidence: as science fiction author Ted Chiang puts it, "In the United States, per-capita G.D.P. has almost doubled since 1980, while the median household income has lagged far behind. That period covers the information-technology revolution. This means that the economic value created by the personal computer and the Internet has mostly served to increase the wealth of the top one per cent of the top one per cent, instead of raising the standard of living for U.S. citizens as a whole."[22]

In this book, we're trying to figure out why the creators of dangerous technologies can be rushing to develop them despite their dangers, and Andreesen's rosy mindset goes a long way toward explaining that. He refers to concerns about unintended consequences of A.I. as a "moral panic," adding that "the public conversation about AI is presently shot through with hysterical fear and paranoia." He argues that, "Big AI companies should be allowed to build AI as fast and aggressively as they can."

"Today," Andreesen concludes, "growing legions of engineers are working to make AI a reality, against a wall of fearmongering and doomerism that is attempting to paint them as reckless villains. I do not believe they are reckless or villains. They are heroes, every one. My firm and I are thrilled to back as many of them as we can, and we will stand alongside them and their work 100%."

Here's a question I think we should be asking here: if the public can be convinced that A.I. is the solution to humanity's biggest problems, how will that boost the market value of the companies that investors such as Andreesen are backing?

Chapter 9

Up to the Stars: Space Colonization and Messaging E.T.s

Once the rockets are up, who cares where they come down?
"That's not my department," says Werner Von Braun.
  —songwriter Tom Lehrer[1]

As I've argued in the last two chapters, how we make technologies is profoundly shaped by how we think of them. We've examined, for example, how extreme techno-optimism has spurred the mistaken belief that providing people with easier means of networking would necessarily strengthen democracies, and we've looked at how this over-optimism—and inevitabilism and solutionism—are motivating today's creators of A.I.

Now let's explore a couple of test cases that demonstrate how these notions might be affecting the creators of two other technologies.

$\sim$

Most people feel a need for meaning and purpose in their lives.[2] Many workers have a modest sense of the importance of their labor: they're content to know that they're providing some useful service, and can feed and house their families. In Silicon Valley, however, many company leaders and employees seem to find it necessary to believe they're engaged on grand missions that will radically improve the lives of all humanity.

After Natalie Goldberg attended that tech lunch I mentioned earlier, she said that she "was surprised to meet the software engineers, who turned out to be fresh, idealistic, and enthusiastic. They said things we used to say as hippies, only they substituted the word *technology* for the word *love*. 'This technology program will change the world, will make it a better place,' intoned one young man."[3]

Many founders recognize the motivational power of that hunger for a great purpose, and they often go out of their way to create a mythos that will harness it. If you watch a documentary about just about any famous tech CEO, you'll see them rallying their employees, promising them glory like Shakespeare's Henry V exhorting his troops before the battle of Agincourt. ("We few, we happy few, we band of brothers!"[4])

Former product manager Antonio Garcia Martinez recalled that, "All the early Facebook employees have their story of the moment when they saw the light, and realized that Facebook wasn't some measly social network like MySpace, but a dream of a different human experience."[5] He added that, "Facebook is full of true believers who really, really, really are not doing it for the money, and really, really will not stop until every man, woman, and child on earth is staring into a blue-framed window with a Facebook logo."[6]

You may well ask, what's wrong with founders espousing noble goals, and their employees being idealistic? Nothing at all— and thousands of tech workers are certainly putting their talents to positive use, ranging from those who help develop life-saving vaccines to those inventing green technologies in hopes of saving our planet. I think of an Industrial Design student at the college where I teach who worked hard to create an inexpensive, easy-to-use home blood test[a] for the millions of women in developing countries who suffer from anemia, and another who designed an accessible kitchen for people in wheelchairs. That kind of socially beneficial zeal is great, and if top dogs want to fire up their employees with a desire to do good, all power to them.

All too often, though, tech leaders become too confident in their own wisdom and powers. As Paulina Borsook put it, "In epics, the protagonists are, by definition, more important to the gods and to men than random mortals could be, which is exactly how people participating in high tech culture post-1992 came to see themselves."[7]

Psychologist Katy Cook recalled meeting a woman at a social media firm who spoke of "the industry's growing arrogance as stemming from a belief that no problem existed that tech could not solve. Such conceit becomes problematic, she explained, when lessons that could be learned from other industries, the past, or the experiences of others are ignored, which might potentially make the products and services of the industry better, safer, or more ethically informed. When I asked why this attitude was so prevalent, the woman described a systemic belief, particularly among executives, which held that those in the industry were the smartest and best suited to solve the problems they were tasked with, and therefore couldn't 'really learn anything from anyone else'."[8]

Cook says that such hubris may actually have a neurological component. She refers to the work of Dacher Keltner, a professor of psychology at the University of California, Berkeley, who "explains that as our sense of power increases, activity in regions of

---

[a] A real one; not a Theranos-style imaginary one!

the orbito-frontal lobe decreases, leading those in positions of power to 'stop attending carefully to what other people think,' become 'more impulsive, less risk-aware, and, crucially, less adept at seeing things from other people's point of view'."[9]

Tech journalist Kara Swisher has spent a lot of time in the company of Mark Zuckerberg, and she has observed that one of his employees might warn him against a particular course of action, but then "he says, 'Forget it. I know best. I remember when I was told not to do this, and I did it, and I won'."[10]

Our new Wheel factor: Unrestrained hubris, especially when manifested in company leaders.

That kind of overbearing presumption has been very apparent recently within Silicon Valley and Washington, D.C., but for the past two decades it has also been playing out in a very different arena, as a trio of billionaires—Elon Musk, Jeff Bezos, and Richard Branson—have competed to launch rockets into space.[b]

You have probably heard three main public reactions to this use of very sophisticated technologies.

The first, a cheer from techno-optimists, proclaims *Hurrah! We're finally getting back to fulfilling humanity's destiny of Progress, ascending toward a glorious future among the stars!*

The second says that the moon—and space itself—are so profoundly barren and unpleasant for humans that it doesn't make sense to spend so much money and effort to establish colonies there. As for Mars, it would seem bad enough to have to live in profound isolation[11] on a sterile planet with extreme tempera-

---

[b] Bezos founded his Blue Origin company in 2000, Musk created SpaceX in 2002, and Branson started Virgin Galactic in 2004.

tures,[c] bombarded with dangerous radiation,[12] looking out only on stark red vistas, without considering that, as journalist Joshua Rothman points out, "NASA missions have found perchlorates in the Martian soil—chemicals that are toxic to humans in even small quantities. Martian dust is so pervasive that avoiding contact with the poison may be impossible."[13] Even so, determined to convince us that space expansionism is a great idea, our trio of tech billionaires are fond of P.R. stunts like sending "ordinary citizens" and the decidedly not-ordinary William Shatner into space.[d]

The third reaction is that these tycoons are just stroking their own egos, and ought to be spending their money to help solve our problems here on Earth.

Enter Daniel Deudney, a professor of political science and international relations at Johns Hopkins University, who offers a fourth reaction that you have likely *not* heard. In a surprising 2020 book called *Dark Skies: Space Expansionism, Planetary Geopolitics, & the Ends of Humanity*, he made a fresh and compelling argument that we may be overlooking mammoth risks in our rush to expand our civilization into space.

When I reached out to Deudney, we spoke via videoconferencing, which seemed oddly appropriate, considering that it was the preferred method of communication on the bridge of the starship *Enterprise*.

We're all familiar with the rhetoric that our ventures into outer

---

[c] The average is about minus 80 degrees Fahrenheit.

[d] Shatner later wrote that he expected to feel great awe during this view of the cosmos, but that instead, "I saw a cold, dark, black emptiness." His main feeling from the trip was profound grief at what we humans are doing to our home planet. (https://www.cnn.com/2022/10/10/business/william-shatner-new-book-boldly-go-scn/index.html)

space and onto other planets represent the best of the human spirit, but as a longtime student of history, Deudney has had plenty of opportunities to examine what tends to happen when humans enter and colonize new frontiers. He's concerned that we might also bring the *worst* aspects of human behavior into space.

The professor notes that we tend to think of our prior space history in terms of its positive sides—the inspiring missions into orbit, to the moon, and to Mars; the cooperation we have seen in the International Space Station—but he says that we tend to ignore the fact that "the dominant space activity" has actually been "weaponization and militarization."[14] He points to the thousands of intercontinental missiles we have built that are currently poised to use space as a fast delivery route for potentially world-annihilating nuclear weapons.

Despite their obvious aggressive potential, we tend to speak of such missiles as a defensive deterrent, and for decades after their development world governments avoided making space a potential battlefield, but that's rapidly changing. Professor Deudney has proved quite prescient on this score: in 2024, the *New York Times*[15] and *CNN*[16] reported that America, Russia, and China were increasingly locked into yet another new arms race, secretly spending hundreds of millions of dollars a year to shift from developing technologies that can defend against attacks in space to creating systems with *offensive* capabilities, employing such techniques as cyberattacks, laser weapons, and high-powered microwaves.

Deudney believes the situation might get even more precarious if we try to expand human civilization beyond Earth's gravity, arguing that, "The propositions of geopolitics applied to solar orbital space" might point to outcomes including "highly violent war, extreme levels of oppression, and the eventual extinction of humanity,"[17] which he refers to as "astrocide." We may think of space as a radically free realm, but Deudney believes that this is not necessarily a good thing: he explains that "frontiers are lawless and commonly very violent, property rights aren't established, and there are incentives to expand at the expense of the others."[18]

He maintains that the region of space directly surrounding the Earth could be a particular danger zone, since any nation gaining control of it would be in a unique position to threaten and dominate the other countries down on *terra firma*. Large floating colonies could be used as platforms that would make it possible for missiles to speed through space and hit their targets even faster than they can now (cutting travel time from about half an hour to under ten minutes). This could jeopardize our current fragile nuclear truce, because any country that could deliver its missiles first would have a huge strategic advantage.

Moving farther out, our future activities would likely include extracting valuable metals from asteroids. Throughout human history, mining has been a grueling, dangerous, exploitative, and politically fraught endeavor. Lewis Mumford wrote that it "was the key industry that furnished the sinews of war and increased the metallic contents of the original capital hoard, the war chest; on the other hand, it furthered the industrialization of arms, and enriched the financier by both processes."[19] Tech researcher Kate Crawford argues that, "Since antiquity, the business of mining has only been profitable because it does not have to account for its true costs: including environmental damage, the illness and death of miners, and the loss to communities it displaces." It was, she says, "the 'move fast and break things' of a different time."[20]

Deudney notes that "the space advocates say, well, there's so much resources, there'll be enough for everybody," but he says that this is probably inaccurate. "We have to assume that the same types of dynamics about resource conflict that have characterized terrestrial history will continue."[21]

While frontiers have tended to be anarchic places, he foresees that colonies in space or on other planets might also have the opposite problem: confined terrestrial vessels such as ships and submarines tend to be run with absolute authority by their captains, and he's concerned that space settlements might tend toward authoritarian, or even totalitarian rule.

Amid these dangers, the professor does note a potential bright side. "Unlike the colonization of remote parts of the Earth, the colonization of space is not going to involve imperialism in the sense of conquering and displacing indigenous populations."

(Unless, I would add, we run into a weaker or more peaceable group of aliens.)

Like the aforementioned tech tycoons, Deudney was enthralled in his childhood by science-fiction tales of space travel, yet he shies away from techno-optimism. When I asked why, he mentioned two other powerful early influences.

"When I was young, in the 1960s, my father worked for Thiokol Chemical Corporation—they basically invented solid rocket fuel. And they [were going to get a] contract with NASA to build a Saturn-grade solid rocket booster. And so they built this facility in South Georgia, which is where we lived, where there's all these marshes—Cumberland, the Golden Isles. My father was involved in hiring all the people that worked at this facility. Ultimately, they didn't get the contract, and after that, we moved away. In 1971, there was this horrific industrial accident. The entire facility went poof, and all these people were just incinerated,[e] and all of these structures were obliterated."

Another formative experience was growing up hiking and camping in that big wilderness region, with "one of the richest

---

[e] 29 people were killed and 50 were injured. It was one of the deadliest industrial accidents in U.S. history.

estuaries in the world," and then—in the 70s—seeing the Navy build the East Coast Trident Submarine base there. "They went in and dredged a big part of it and made channels and stuff. So right there in the heart of nature, we've got one of the largest nuclear weapons repositories on the planet."

Now, as a grown man who has written about the potential dangers of space expansionism, Deudney has found himself a lone voice in another wilderness. He told me that he had expected that his book would get strong media coverage and generate a healthy public dialogue, but he was disappointed to find that it received surprisingly little attention. Bemused, he speculated about some of the journalists who write about our space ventures. "They're advocates, they're enthusiasts. They could be working in an industry publicity department. They're not at all interested in any questioning of the larger narrative."[22] The professor sighed. "We really don't have any significant organized opposition to the core expansionist vision."[23,f]

I can only agree with his assessment: when it comes to technologies like social media or A.I., our society has become increasingly concerned about negative consequences, but in regard to space we still seem eager to buy into the techno-optimist worldview. This blinkered approach applies not only to journalists and the general public, but to the creators of space-oriented technologies themselves. As Deudney put it, "Those who think most about space do not commonly question whether space activities will be

-----

[f]   The professor *did* get some recognition in a recent book that presents another rare challenge to space colonization over-optimism. In *A City on Mars: Can We Settle Space, Should We Settle Space, and Have We Really Thought This Through?* researchers Kelly and Zach Weinersmith present compelling evidence that today's boosters of expansionism, such as Elon Musk, are overlooking a slew of reasons why settlements would present enormous physical, medical, psychological, legal, and political challenges. (New York: Penguin Press, 2023)

positive in their consequences because their primary focus is making space activities more feasible."[24]

(I would point out that this statement could apply to the people creating just about all of the technologies we are examining in this book. All too often, their main question is not <u>*Should*</u> *we do this?* but <u>*How*</u> *can we do this?*—and *How can we do it first and dominate the field?*)

The professor points out that it's particularly difficult to criticize space expansionism because it's bound up in a number of traditional conceptual and ideological propositions. For example, "There's been this deeply rooted and completely dominant tendency to view going *up* as being intrinsically good. It's written into our language as well. Take the word 'superior': it means both *above* spatially and *better*. Or we talk about the wealthier social class as the "upper" one. States that are becoming more powerful and influential are 'rising powers.'"[25]

The connotations go beyond language. Religions tend to visualize their heavens as *up*. As Deudney puts it, "The long historical association of spatial elevation with moral and religious improvement gives this religiously infused space expansionist worldview an intuitive credibility. This narrative imparts to its believers a strong faith that their small and struggling steps inevitably contribute to something of vast cosmic import."[26]

As I've just mentioned, tech founders love imbuing their companies with that kind of mythos. Deudney notes that the idea of space expansionism "acquires an ethical imperative of the highest caliber,"[27] including—in the minds of people like Elon Musk—the very salvation of humanity.

(Speaking of Musk, he has shot thousands of Starlink internet communications satellites up into near-Earth orbit, blanketing the globe, and his SpaceX company now dominates this industry. To his credit, Musk provided a great deal of technical equipment and support to help Ukraine protect itself from the 2022 Russian invasion, but that country's top brass became nervous about relying on a very erratic tech mogul for such an essential service—as did a number of world leaders. They were worried that Musk could unilaterally decide to pull the plug on connections to his network at any moment—and in fact, he did so, later that year, when Ukraine tried to attack the Russian Navy in Crimea.[28] As cybersecurity expert Dmitri Alperovitch put it, "This is not just one company, but one person. You are completely beholden to his whims and desires."[29])

Deudney maintains that the space expansionist venture is also intimately entangled with an economic and political philosophy. "Space advocates herald these initiatives as a decisive break from previous government-led space efforts and have high expectations of dramatically lower costs and large new markets. They increasingly speak the language of American libertarianism, with its conviction that government is primarily a problem and progress is assured only when market forces are fully unleashed."[30] For these ultra-wealthy boosters, "space expansionism offers an escape from government regulation and taxation." They claim that "unleashed capitalism offers the fastest way to make space happen in a big way."

Once again, Deudney's perspective on our human past makes him question the notion that this is a good idea. "Whatever the virtues of market exchange within a well-defined system of property, capitalism—and the state-system—have been historically atrocious in making a fair 'original distribution' when a *res nullius* (a thing of no one) is converted into property or sovereign territory. Will the further opening of space intensify the steep and growing stratification of wealth?"[31]

There can be a great difference between technological smarts and moral wisdom, and Deudney is acutely aware of this. "And so we ask the question, have our capacities to govern grown as rapidly as our capacities to destroy both ourselves through violence and the planet through destructive practices? You clearly have to say, No, there's a gap.' On the one hand, we've got this rapidly rising curve of technological enablement, but then you ask, how is our capacity to restrain on a global scale comprehensive with regard to these technologies, and it's still primitive."[32] Deudney believes that our situation was well described by American biologist E. O. Wilson, who said that, "The real problem of humanity is the following: we have Pleistocene emotions, medieval institutions, and god-like technology."[33]

As we spoke, he chuckled ruefully as he mentioned a perfect symbol of extreme techno-optimism. "I always remember this cartoon from *The New Yorker*. It's got a cliff and then it's got a row of lemmings, this legendary rodent that supposedly throws itself into the ocean. And there's a row of them falling off the cliff to sure demise, and then there's a dotted line with images of lemmings going up, and the caption is 'What lemmings believe'."

This may all seem quite negative, but Deudney hastens to point out that—though he's opposed to space colonization—he's not at all opposed to space *exploration*, in the interest of furthering our scientific knowledge.

And he argues that there's one more fundamental upside to the issue: "Unlike many other threats humanity faces, addressing those created by ambitious space expansion is now extremely simple: just say no."[34]

Now, I realize that for many people, Deudney's arguments may appear extreme. I can certainly understand that: I was raised on the mythos of the astronauts too, and have always been moved by the incredible courage of the men who ventured all the way to the Moon, with relatively primitive technology, in a little tin can.

I don't want to rain on that parade, but if you read Deudney's book, you'll see that it's a very well thought-out one, backed up with a great deal of evidence—he deserves a fair hearing. If we move ahead with our current thrust up beyond our home planet, at the very least it would provide an excellent guide to questions we ought to ask along the way.

What are the risks we should be considering, and what are the precautions we ought to take?

We were just examining the risks of human expansion from the Earth into space, but I'll end this chapter by mentioning another threat that could flow in the opposite direction. (But also fueled by extreme optimism and inevitabilism.)

In Chinese science fiction writer Liu Cixin's novel *The Three-Body Problem*, an astrophysicist working at a remote radio telescope receives a transmission coming from extraterrestrials living in another galaxy. For reasons of her own, without consulting anyone else, she sends out a response—and the aliens immediately head toward Earth, with less than benevolent intentions.

Unfortunately, this scenario is not just speculative. In the past century we humans have already beamed signals out into the stars. We have done so unintentionally, via our radio[g] and TV broadcasts, and we have done so on purpose, starting with a 168-second encoded message sent from the Arecibo radio telescope in Puerto Rico on November 16, 1974. This transmission was the idea of astronomer Frank Drake, and within days it had spurred an irate response from Martin Ryle, the Royal Astronomer of England, who argued that we had no idea whether the message might be received by a malevolent alien civilization that could want to wipe us out.[35,h]

---

[g]   They have been traveling into space for more than 100 years, and as a headline in *Discover* magazine put it, "Our Radio Signals Have Now Reached 75 Star Systems That Can See Us Too." (https://www.discovermagazine.com/planet-earth/our-radio-signals-have-now-reached-75-star-systems-that-can-see-us-too)
[h]   The Pioneer 10 space probe had been launched two years earlier, bearing a

Given the fact that Drake's Arecibo message was directed at a star cluster 250 trillion miles away, this might seem like quite a distant concern in time as well space, but recently scientists have discovered possible Earth-like planets that are much closer than that. What might happen if we messaged them?

Funny you should ask.

You have likely heard of S.E.T.I., an international organization dedicated to the Search for Extraterrestrial Intelligence. In 2015, one of its members, an astrobiologist/psychologist named Douglas Vakoch, broke away from that group and founded a San Francisco-based international organization dedicated to M.E.T.I., or *Messaging* Extraterrestrial Intelligence.

Interestingly, the prominent critics of such an attempt have included Stephen Hawking and Elon Musk.[36,i] The latter signed a group statement opposing M.E.T.I., which said that, "A worldwide scientific, political, and humanitarian discussion must occur before any message is sent." Another objector was astronomer David Brin, who has argued that, "Every single case we know of a more technologically advanced culture contacting a less technologically advanced culture resulted at least in pain." He concluded that, "I just don't think anybody should give our children a *fait accompli* based on blithe assumptions and assertions that have been untested and not subjected to critical peer review."[37]

Despite such warnings, in 2017 M.E.T.I. sent a message to the

---

famous gold-anodized plaque that featured nude figures of a man and a woman, but it travels at an extremely sluggish rate compared to Drake's radio waves. And in 1977, two Voyager spacecraft left our solar system bearing actual gold records, one of which featured the song "Johnny B. Goode." I'm a fan of Chuck Berry's music, but the notion that an extraterrestrial civilization would be able to hear it and make the slightest sense of it—or the hit tracks by Bach or Stravinsky—is so farfetched as to seem quite bizarre. In retrospect, this just seems like another P.R. stunt designed to appeal to humans, with no consideration of the potential risks.

i I'll give Musk credit for one thing: he's not easy to predict. At times he behaves like a bratty adolescent; at other times he comes across as the very voice of reason.

red dwarf Luyten's Star, which is just twelve light years away from Earth.[38] If there happens to be an alien civilization near it, we will have alerted them to our presence by 2029. We don't have any international laws or treaties that block such an action, so—like Liu Cixin's lone astrophysicist—one small group of scientists and techies was able to make it completely on their own.

When Steven Johnson asked anthropologist Kathryn Denning, who has debated the M.E.T.I. approach, about her stance on the issue, she replied: "I have to answer that question with a question: Why are you asking me? Why should my opinion matter more than that of a six-year-old girl in Namibia? We both have exactly the same amount at stake, arguably, she more than I, since the odds of being dead before any consequences of transmission occur are probably a bit higher for me, assuming she has access to clean water and decent health care and isn't killed far too young in war."

"In the final analysis," Denning concluded, "it's all about how much risk the people of Earth are willing to tolerate . . . And why exactly should astronomers, cosmologists, physicists, anthropologists, psychologists, sociologists, biologists, sci-fi authors or anyone else (in no particular order), get to decide what those tolerances should be?"[39]

There she's touching on a fundamental question that applies not only to M.E.T.I., but to many other new technologies: why should a few rich private entrepreneurs and their companies get to unilaterally choose to unleash these on our planet?

Wheel factor: an extremely limited pool of people making the decision to develop and employ a new technology.

As another example, let's look at A.I. It's hard to say how many tech professionals are actively building the technology at the moment, but one 2018 report estimated that, globally, there were about 22,000 specialists with the expertise to create machine learning systems, which form the basis of advanced A.I models.[40]

That means that a quarter of a thousandth of one percent of the human race were making the choice to create this technology that would soon sweep the planet and change all of our lives. In the past several years that number of experts has certainly swelled, yet the fact remains that an incredibly small number of humans are deciding that it will be at the center of our human future—and that eight billion others should be forced to assume its risks.

When do the rest of us get our say?

Chapter 10
On Ideology: A Very Strange Icon

"In our time, the great task for libertarians is to find an escape from politics in all its forms—from the totalitarian and fundamentalist catastrophes to the unthinking demos that guides so-called 'social democracy' [...] We are in a deadly race between politics and technology. [...] The fate of our world may depend on the effort of a single person who builds or propagates the machinery of freedom that makes the world safe for capitalism."[1]

—PayPal cofounder and venture capitalist Peter Thiel

BY THIS POINT, AFTER SUCH A LITANY OF PROBLEMS WITH technology, I'm sure you're eager to get to some solutions. I am too, but before we do, I'd like to delve into just two more spokes of our Wheel.

As we have seen, the ways in which technologists operate are affected by the culture in which they work, their psychological makeup, their view of history and progress, and how they think about technology. They're also shaped by particular ideologies.[a]

Speaking of which, what one human being would you guess has provided the most impactful set of ideas and values for Silicon Valley in the digital age?

Is it a founder of Apple, or Microsoft, or Google, or Facebook?

No.

Is it a notable scientist or inventor?

No.

Is it an industrialist, like Henry Ford, an author of corporate management theory, a venture capitalist, or a CEO?

Nope.

I'll give you a hint: this person was not a technologist at all, and she died in 1982, seven years before the invention of the World Wide Web.

As tech journalist Nick Bilton put it, "Perhaps the most influential figure in the industry, after all, isn't Steve Jobs or Sheryl Sandberg, but rather Ayn Rand."[2]

The Russian-born author, best known for her novels *The Fountainhead* and *Atlas Shrugged*, has long been a major influence in the realms of economics and politics in the U.S., from former Fed Chairman Alan Greenspan (a member of her inner circle) to Ronald Reagan to Paul Ryan, and her ideas have also reached deep into the world of tech. Rand biographer Lisa Duggan notes that her fans in Silicon Valley have included tech leaders Steve Jobs, Peter Thiel, Travis Kalanick, and Jeff Bezos, "ad infinitum."[3]

We might well call Rand the patron saint of the Wheel, and the presence of true believers is a prime indication of when we're deep inside the hub of it.

---

[a] *Oxford Languages* defines an ideology as "a system of ideas and ideals, especially one which forms the basis of economic or political theory and policy."

Considering that many of Rand's devotees have been conservatives, in some ways she seems like a very strange icon.

The traditional American conservative is supposedly a Christian, churchgoing, temperate, monogamous, anti-abortion, military hawk. Rand, on the other hand, was born Jewish, and at age thirteen she became an avowed, lifelong atheist.[4] Though married, she had a decades-long affair with one of her most devoted students,[5] and for three decades she was a Benzedrine (speed) addict.[6] She was opposed to the Vietnam War, the military draft, and legal bans on abortion.[7,b]

Lest she start sounding like a liberal, let's note that she also hated F.D.R.,[8] was fervently anti-union,[9] was a friendly witness before Joseph McCarthy's House Un-American Activities Committee,[10] attacked the 1964 Civil Rights Act,[11] was an avid proponent of unregulated capitalism,[12] called homosexuality "immoral" and "disgusting,"[13] and detested the environmental movement.[14] In regard to the European genocide of Native Americans, she said, "Any white person who brings the elements of civilization had the right to take over this country,"[15] and she also declared that she would not vote for a female president.[16] She rejected the concept of altruism, and published a book of essays titled *The Virtue of Selfishness*. (In general, her philosophy is an argument against emotional intelligence.)

I'll mention just one last biographical detail, because it's especially germane to the relationship between tech moguls and

---

b To be clear, I'm not saying that there's anything at all wrong with being Jewish, atheist, anti-war, or in favor of abortion rights. I'm just pointing out that Rand makes an odd poster child for conservative Christian values. (She's almost as odd as Donald Trump—who has said that he identifies with Howard Roark, protagonist of *The Fountainhead*. https://nymag.com/intelligencer/2016/04/trumps-role-model-is-an-ayn-rand-character.html)

government. After a lifetime of excoriating those who rely on social support systems as "parasites," in her final years Rand took Social Security and Medicare benefits.[17]

How could someone with so many odious qualities become revered by so many powerful people?

Often, the love affair begins at an early age. Adolescents often think, *How can these authority figures all around me fail to recognize how smart and special I really am? I'll show them!* This theme of being underappreciated and misunderstood, a central characteristic of Rand's fictional heroes, likely explains her notable appeal for teenagers; one of them was Wendy Liu, who wrote that, "During a particularly bleak stretch of my adolescence—marked by the perilous blend of impeccable grades and lonely lunches—I had fallen in love with Ayn Rand's novel *The Fountainhead*, which I took as confirmation that I would eventually be avenged for my current lack of recognition."[18]

Liu grew out of her youthful Rand-enamored phase, as many people do, but Rand herself seems to have been stuck in a lifelong adolescence—and she also wanted to be avenged. The Bolsheviks had seized her upper-middle-class family's assets in Russia when she was twelve,[19] which meant no more private education for her, no more governess, no more large apartment, and no more fancy vacations, and it sent her father into a deep depression.[20] She went on to nurse a lifelong resentment of government and other collective social institutions.

That aspect of her viewpoint has proved particularly resonant for a slew of tech moguls, who want to believe that they—like the heroes of her novels—are macho, visionary builders and creators beset by ignorant, obstructive government officials, and carping critics, and Socialists who stupidly believe in a common good. According to Bilton, "At their core, Rand's philosophies suggest that it's O.K. to be selfish, greedy, and self-interested, especially in business, and that a win-at-all-costs mentality is just the price of changing the norms of society. As one start-up founder recently told me, 'They should retitle her books *It's O.K. to Be a Sociopath!*' And yet many tech entrepreneurs appear to live by one of Rand's defining mantras: 'The question isn't who is going to let me; it's who is going to stop me'."[21]

Journalist Jonathan Freedland further explains that, "Among these new masters of the universe, the Rand influence is manifest [...] in a single-minded determination to follow a personal vision, regardless of the impact. No wonder the tech companies don't mind destroying, say, the taxi business or the traditional news media. Such concerns are beneath the young, powerful men at the top: even to listen to such concerns would be to betray the singularity of their own pure vision."[22]

Most of us have come to expect and accept that a company's executives will have to operate within strict government guidelines if they want to sell a new pharmaceutical, build an airplane, operate a bank, manage a hospital, or open a power plant. And yet, to many, it seems outrageous that we might impose any controls on a company that wants to release a major new technology. In fact, during the entire internet era, Big Tech has benefited from an almost complete lack of oversight.

Ayn Rand's fingerprints are all over that fact.

Her loathing of regulation has clearly appealed to many tech moguls. Like her, they love to claim that government is an ignorant, regressive force that interferes with the bold march of innovation and Progress.[23] Barack Obama summed up this attitude he had often heard in his encounters with them: "The last thing we want is a bunch of bureaucrats slowing us down as we chase the unicorn[c] out there."[24]

In Silicon Valley, this antipathy has long been extreme. In its early days, the sentiment was proclaimed by its conservative libertarians,[d] but also by its hippies. In 1996, tech activist (and sometime Grateful Dead lyricist) John Perry Barlow penned a now-notorious "Declaration of Independence of Cyberspace" in which he proclaimed, *"Governments of the Industrial World, you weary giants of flesh and steel, I come from Cyberspace, the new home of Mind. On behalf of the future, I ask you of the past to leave us alone. You are not welcome among us. You have no sovereignty where we gather. [...] You have no moral right to rule us nor do you possess any methods of enforcement we have true reason to fear."*[25] Barlow published his manifesto in *Wired* magazine,[26,e] where he was a frequent contributor. The publication was particularly influential in shaping the Valley's ideology; its founder and publisher Louis Rossetto was a Rand fan[27] and a hardcore libertarian, and the early issues strongly reflected his anti-government bent.[28]

Earlier in the last century, though, government oversight had *not* been so frowned upon in the U.S.

In regard to communications technologies, legislators passed the Radio Act in 1927, which stated that broadcasters needed to obtain licenses, and encouraged their companies to act in the public interest. In 1934 they created the Federal Communications

---

[c] "Unicorn" is a term for a venture capital-backed private tech company valued at over one billion dollars.

[d] Although she is often seen as a libertarian, Rand didn't identify herself that way; she actually called its proponents "a monstrous, disgusting bunch of people." I think she was just angry that they had rebranded her ideas. (http://aynrandlexicon.com/ayn-rand-ideas/ayn-rand-q-on-a-on-libertarianism.html) In any case, her revulsion toward government regulation, labor unions, social support programs, and collective political action strongly influenced the libertarian ethos.

[e] *Wired*'s founding editor was Kevin Kelly, who has said that he "blew a gasket" and radically changed the direction of his life after he read Ayn Rand's *The Fountainhead*. (https://www.afr.com/politics/totally-wired-20010309-j781r)

Commission, and in 1949 it established what became known as the Fairness Doctrine, which mandated balanced coverage of political points of view.[29] The government also moved to regulate other aspects of what went out over the airwaves, including enacting a series of laws that prohibited deceptive marketing and advertising.

In 1963, alarmed by the great dangers posed by nuclear weapons technology, the nation managed to join with other countries to limit the arms race in space by creating the Limited Test Ban Treaty, and it signed the Outer Space Treaty of 1967. And the Nixon/Ford years were a boom time for government regulation: Wendell Wallach notes that "nearly all the major U.S. environmental laws were enacted between 1969-1976."[30]

But anti-government activists soon had their way. The Vietnam War soured the public's opinion of its leaders, and Watergate further damaged their trust. The time seemed ripe for a rebellion against the notion of benevolent government, and on the political Right, the charge was led by devotees of Ayn Rand.

Langdon Winner argued that Ronald Reagan's election in 1980 "signaled the end of a favorable climate of opinion for public discussions of energy policy, energy conservation, and alternative energy technologies that had characterized the presidency of Jimmy Carter. 'Let the market work,' was now to be the central premise of government policy. During the era of Reaganomics, most planning of nonmilitary technologies moved back to the private sector, back into the hands of business corporations and away from public scrutiny."[31]

President Reagan, backed by economists such as Milton Friedman and Alan Greenspan, soon tore into government regulations like a hurricane, in ways that would have major impacts on the emerging Silicon Valley. Friedman declared that in a free society, "there is one and only one social responsibility of business—to use its resources and engage in activities designed to increase its profits."[32,f]

During the early years of the digital era, the movement to protect its businesses from government oversight had two other especially powerful backers. Charles Koch, another major Rand fan,[g] is the owner of Koch Industries, the second-largest privately held company in the U.S. (He co-owned it with his brother David, who died in 2019.) In his book *Move Fast and Break Things: How Facebook, Google, and Amazon Cornered Culture and Undermined Democracy*, scholar Jonathan Taplin argues that, "The Kochs are important because they financed the rise of the libertarian political framework that Peter Thiel, Larry Page, Jeff Bezos, and Mark Zuckerberg used to get rich. Without the political protection of the Koch network, none of the Internet empires would exist at its current scale."[33]

Brookings Institute researcher Darrell West has written about how, in 1995, Newt Gingrich shut down the U.S. government's Office of Technology Assessment, which had been in operation since 1974 and was designed to provide Congress with impartial analysis of technology and science issues.[34] "It was time to eliminate that agency, he claimed, because it was ineffective and major tech decisions should be made by the private sector. Members of his party agreed with his stance and voted to get rid of the tech research and policy shop. Ironically, legislators killed that agency just as the internet was unfolding."

"At the very time when Congress needed advice on how to manage the consequences of powerful new technologies," West concluded, "an agency that could have been helpful in assessing

---

[f]   He qualified this by adding "so long as it stays within the rules of the game, which is to say, engages in open and free competition without deception fraud," but the first part of the statement remains startlingly Randian.

[g]   The Ayn Rand Institute is a partner organization of the Charles Koch Institute. (https://www.sourcewatch.org/index.php/Ayn_Rand_Institute)

new tools was mothballed. Nonpartisan experts could have offered their views on how to handle privacy, security, bias, transparency, and human safety. Instead, policymakers lacked any systematic federal contribution to public discussions regarding digital technology's upsides and downsides, or the role government should play in tech's ongoing development. It was stunningly poor timing on the part of Gingrich and his fellow legislators at a crucial point in the digital revolution."

Even when governments take a more hands-on approach, there's often quite an interval (sometimes decades) between the time when entrepreneurs introduce a new technology and when regulators catch up with it.[h] For a tech company eager to dominate, that lag time can offer major opportunities, and the internet provided a prime example. As Shoshana Zuboff has argued, "A key element of Google's freedom strategy was its ability to discern, construct, and stake its claim to unprecedented social territories that were not yet subject to law."[35]

Some tech moguls are so opposed to government oversight that they have even advocated for physically moving tech companies out of its reach. Peter Thiel has argued that this would be one benefit of space expansionism, but noting that this might take a while, he once focused considerable effort on *seasteading*: creating floating colonies here on the Earth's oceans that would be outside the jurisdiction of terrestrial governments.[36]

A similar antipathy is at the heart of a 2021 book called *Where Is My Flying Car?*[37] written by J. Storrs Hall, a scientist who has

---

[h] Here are some intervals between invention of technologies and the first major federal regulation: for railroads, 60 years; for cars, 70 years; for telephones, 30 years. (https://www.nytimes.com/2023/08/24/upshot/artificial-intelligence-regulation.html)

worked in the fields of nanotech and A.I. The book is predicated on the odd argument that though science fiction authors once promised us flying cars, we were robbed of this somehow-essential technology by a number of dastardly enemies of innovation and Progress during the last half century, a period which Hall refers to as "the Great Strangulation."[i] He attributes this mostly to government regulation, and his language here is all too familiar: "The stunning progress of the Industrial Revolution from the 1700s to the 1970s wasn't a miracle but the result of a large enough percent of enough generations being determined to be greater than the last and innovating their way, through technology and science, toward being greater in fact. They were able to do so in part because the entangling brambles of regulation that snag us at every turn largely weren't there."[38,j]

Hall is hardly a fan of the last century's countercultures, but his attitude toward government and regulation reminded me of what journalist Paulina Borsook once wrote about the *cypherpunks* of the 1990s, whom she described as "radical pro-privacy computer activists." She observed that, "Cypherpunkery at its most extreme sees government as a monoculture, peopled only by the unprincipled, the dull-witted, the corrupt, and the power-trip-

---

[i] Hall makes all sorts of strange assertions, including an argument that this lull in technology was partially caused by too much higher education, and he calls environmentalism "an apocalyptic nature cult," labels global warming a "scare story," and claims that it is the scientific consensus that "the impact of climate change will be small." He writes of conventional cars as if they have been an absolute benefit to the world, and calls public transit "worthless." Amazingly, though he does an exhaustive and impressive analysis of what would make flying cars technologically feasible, he never once stops to wonder what it might be like to have our skies thronged with thousands of noisy little vehicles, or where city dwellers would park them, or why it might not be a good thing (as he claims) if these new cars would enable human beings to spread out across even more of the surface of the Earth.

[j] Presumably, even more Progress could have been made if we hadn't stopped letting employers fill rivers with arsenic and lead, making laborers toil in twelve-hour shifts six days a week, and forcing small children to work in mills and mines.

ping. It is an angry adolescent's view of all authority as the Pig Parent, uniformly cretinous and bad and oppressive and seeking domination—rather than as a complex institution with a variety of good and bad actors with different proclivities, drives, and intelligences."[39]

Science fiction author Robert A. Heinlein, cited as a major inspiration by space entrepreneurs Elon Musk and Jeff Bezos, was a self-described radical libertarian.[40] In 1966 he published *The Moon Is a Harsh Mistress*, a novel about Moon-based colonists who engage in armed rebellion against an annoying, interfering Earth government. I have to wonder if—just three years later—when humans first landed on the Moon, the author saw any irony in the fact that the achievement was primarily funded by the U.S. government, through NASA.

That irony has also played out here on Earth, in real-life Silicon Valley. Like Ayn Rand in her final years, tech moguls have long complained about government interference while relying on its largesse. In fact, the whole tech industry came into being largely due to massive government help. This includes the publicly funded infrastructure that its workers rely on, from the electric grid that powers their computers to the streets they drive on get to work, from the state colleges that gave many of them their training to the funding for much of their research.

Economist Mariana Mazzucato notes that, "All the most important technologies that have driven growth—what economists call *general purpose technologies*—trace their funding back to government. [...] These include aviation technologies; space technologies; semiconductors; the Internet; nuclear power; and nanotechnology."[41] For good measure, she adds that, "All basic technologies that make our mobile phones 'smart' can be traced back to governmental initiative and funding. Just a few: microprocessors; RAM memory; hard disk drives; liquid-crystal displays; lithium batteries; [...] cellular technology and networks; global positioning system (GPS); multi-touch screens."[42]

Paulina Borsook grew up in Southern California, and she had fond memories of "Cal Tech/Jet Propulsion Lab/Southern California aerospace guys." In that earlier version of Silicon Valley, she recalled, "the engineers and scientists more likely than not shared a vaguely New Deal mentality. They were of a generation that had seen what good things the government could do, from winning World War II to putting a man on the moon."[43]

There's a big distance between that mindset and the current one in the world of big tech. In December of 2021, Elon Musk told a *Wall Street Journal* conference that "the government should just try to get out of the way," and, "I'm literally saying get rid of all subsidies" [44]—which seems quite odd given the fact that, just several months later, journalists reported that his companies, including Tesla and SpaceX, had "received more than $7 billion in government contracts alone and billions more in tax breaks, loans and other subsidies."[45]

By 2025, that figure had swelled to $38 billion.[46]

If we want to understand Musk, it's important to note that he was not just inspired by Rand and Heinlein; as journalist Jill Lepore has pointed out,[47] some of his ideas were heavily influenced by his maternal grandfather Joshua Haldeman, who was one leader of a bizarre political movement, surprisingly popular in the early 1930s, called Technocracy. As Lepore explains it, "Leading technocrats proposed replacing democratically elected officials and civil servants—indeed, all of government—with an army of scientists and engineers under what they call a technate."[k] Under

---

[k] In this society, humans would have been given numbers instead of names (which may explain why Musk actually named one of his children X AE A-12.)

this regime, voting would be "largely dispensed with," most government services would be eliminated, and "90 percent of the courts could be abolished."

In 2025 Musk got the opportunity to put some of those technocratic fantasies into practice—this may explain why he took that disastrous wrecking ball to a whole slew of U.S. institutions with his hideously misnamed "Department of Government Efficiency."

Again, I don't want to tar everyone in tech with the same brush. Though a number of its leaders have clearly been influenced by the Randian spirit of megalomania and selfishness, in 2019 Stanford University researchers found that the Valley's tech workers were actually politically liberal in most respects: in favor of social services for the poor and high taxes on the rich, and concerned about environmental protection and income inequality.[48] On the other hand, they also found a continuing strain of resistance to government oversight and regulation, leading them to come up with the new label *liberaltarian.*

But as I mentioned earlier, the recent techlash spurred a surprising development: in its wake, some Silicon Valley leaders began showing a new openness to government oversight. In a Congressional hearing about social media in 2018, Mark Zuckerberg testified that "it is inevitable that there will need to be some regulation,"[49] and in a 2023 Senate hearing on A.I., Sam Altman testified that, "We think regulatory intervention by governments

---

And speaking of explanations for odd present-day phenomena, one has to wonder if Donald Trump's seemingly inexplicable fixation on annexing Canada and Greenland (not to mention the Panama Canal) was actually inspired, via Musk, by the technocrats' notion that the U.S. and those places should be united in a "Technate of America." (https://digital.library.cornell.edu/catalog/ss:34227574)

will be critical to mitigate the risks of increasingly powerful models."[50]

The Biden administration started working with big A.I. companies to increase safety testing and develop safety standards, but the election of 2024 reversed this positive trend. As a 2025 *New York Times* headline put it, "Emboldened by Trump, A.I Companies Lobby for Fewer Rules."[51] Big Tech saw a promising ally in the new administration, especially since Vice President JD Vance had strong ties to Silicon Valley, having been mentored and financially backed by anti-regulation mogul Peter Thiel.[52] In a 2025 speech, Vance stated that, "The AI future is not going to be won by hand-wringing about safety."[53] [1]

Ayn Rand would have been proud.

---

[1]   Vance, Musk, Marc Andreesen, Jeff Bezos, and Peter Thiel are not just fans of Ayn Rand—they're also major enthusiasts of J.R.R. Tolkien and his *Lord of the Rings* trilogy—despite the fact that Tolkien himself was not a fan of the machine, which he saw as having a corrupting power. "If there is any contemporary reference in my story at all," he once wrote, "it is to what seems to me the most widespread assumption of our time: that if a thing can be done, it must be done. That seems to me wholly false." (https://www.nytimes.com/2025/05/23/books/tolkien-musk-thiel-silicon-valley.html)

Chapter 11
The Elephants In The Room: Venture Capitalists, Monopolists, And Hyper-Growth

"I do not believe the platforms are causing harm on purpose. The harm is a by-product of hyper-focused business strategies that failed to anticipate negative side effects. These are really smart people operating in a culture that sees the world through the narrow lenses of business metrics and code.[1]

—Roger McNamee

IF ONLY WE COULD DIVERSIFY THE WORLD OF TECH, AND increase its emotional intelligence. If only we could reduce surveillance of users and violations of our privacy, and eliminate extreme algorithms, and replace the Randian ideology that permeates Silicon Valley with a healthier, more humane philosophy.

Given what we have considered in the past few chapters, you might think that if we could achieve those goals, we'd be well on our way to improvement. And so we would—but we'd still face another major obstacle. The industry would remain snared in a business model that tends to send its creations in bad directions.

We often seem to think of digital media and the internet as unprecedented technologies, but—like Tim Wu—Douglas Rushkoff argues that their fundamental economic model is hardly new. He has described how, centuries ago, business tended to take place in bazaars, in a "peer-to-peer economy where laborers and small merchants could engage in barter and exchange,"[2] but then the earliest corporations managed to wrest away those workers' power and profits.

Rushkoff told me that, "Once the lords of feudalism decided to disable the growing local economies of merchants by forcing them to use extractive central currency, or to work for chartered monopolies instead of for themselves, we put in place a different kind of economy that was based on extraction and domination. (Rather than even Adam Smith-like, bottom-up competition.) So you get the British East India Trading Company being modeled by Walmart or Monsanto. And Walmart and Monsanto being modeled by Uber and Amazon. It's the same basic game plan where you find a working market, dominate it, and monopolize it. You externalize the real costs, whether it's pollution or labor or autonomy, while you extract the value."[3]

But while today's tech companies use tactics developed over centuries of capitalist practice, there *are* new aspects to how they operate.

Funding from venture capital, a new means of financing companies that was developed after World War II, is one of the things that distinguishes modern tech companies from their forebears. This model is different because, unlike bankers, who lend out cash and expect repayment, venture capitalists don't make loans; they give money outright to fledgling businesses, in exchange for a stake. And while bankers take great care to lend only to companies they consider likely to repay them, V.C.s know that a number of their startups will probably fail. That might seem like an odd plan for a successful business, but they rely on occasional smash hits (a Digital Equipment Corporation, a Microsoft, an Apple) that will more than make up for the losses. The companies that V.C.s fund are often those they deem capable of rapid, exponential growth.

Traditional companies need to generate revenue and profits very early on. Fat with V.C. funding, though, a tech company can start out offering its service for such a low price—or *no* price—that it actually *loses* money for its first several years. It can scale up—globally—very cheaply if it relies on the digital internet, rather than physical factories or stores. Once it puts its basic software in place, it can generate an exponential increase in the number of users with little corresponding rise in costs, including the cost of labor.

Corporations like General Electric or General Motors used to take great pride in the thousands of jobs their businesses provided, bolstering a broad American middle class, but Jonathan Taplin points out that even though tech companies now make up about twenty percent of the S&P's top 500, they employ only three percent of American workers.[4] Digital technologies also make it easier for companies to export jobs overseas, where workers can be paid far less.

Another thing that distinguishes V.C.s from traditional investors is that they aren't necessarily interested in the long-term viability and success of the companies they're funding; sometimes, their main goal is simply to help grow the companies quickly, then sell them as soon as they become highly valuable. (This is known as "taking an exit.") As Rushkoff explains it, "creating a company for acquisition or IPO is different from building a profitable enterprise; it's about building a *sellable* enterprise. Startups are not trying to earn revenue (which is a liability); they are setting themselves up to win more capital."[5]

To some V.C.s, it makes little difference whether their startups are even creating anything useful. In 2020, one founder (who wished to remain anonymous) told journalists Ben Tarnoff and Moira Weigel that, "I think what really shifted my thinking about my success was observing that most of the startups that are getting crazy amounts of venture capital aren't solving interesting problems. There is a lot of money going into squeezing more ad dollars out of users, and getting more attention and eyeballs. [...] I also felt like people were becoming founders and investors not because they wanted to solve problems that would help humanity, but because they wanted to be in the Silicon Valley scene."[6]

This business model has a huge impact on what kinds of technologies startup founders create. Often, V.C.s decide that the way to build a sellable company is to ravage an existing industry. Antonio Garcia Martinez has said that "technology entrepreneurs are society's chaos monkeys, pulling the plug on everything from taxi medallions (Uber) to traditional hotels (Airbnb) to dating (Tinder). One industry after another is simply knocked out via venture-backed entrepreneurial daring and hastily shipped software."[7]

The leaders of these new tech companies didn't just want to see their companies grow; they wanted to take over their whole markets.

Mark Zuckerberg liked to galvanize his employees at early Facebook meetings by leading them in chants of "Dominate! Dominate!"[8] One reason he was able to succeed at doing so was explained by Jaron Lanier in his excellent book *You Are Not a Gadget*, when he argued that, "Digital network architectures naturally incubate monopolies."[9]

A drive for monopoly power is another common element of the Wheel.

The Frankenstein Fix

Of course, though the internet and other digital technologies have provided fresh ways to achieve it, the desire to dominate is not new. At the end of the 19th century, a few industrialists gained massive control over businesses such as railroads,[a] oil fields, steel manufacturing, and telephone networks. In that largely unregulated era, people like John D. Rockefeller and Andrew Carnegie were able to corner markets, force competitors out of business, and amass staggering fortunes. The "robber barons" had free rein for decades until public officials passed a number of restraining laws, such as the Sherman Antitrust Act of 1890. Antitrust lawyers and politicians forced the breakup of some of the biggest monopolies, such as Standard Oil and AT&T. The 20th century saw rough times in its first half, with two world wars sandwiched around the Great Depression, but its middle years, heavily regulated, with high corporate taxes, saw rising middle-class wages, thriving industrial innovation, and strong economic growth.

Here we come to another historical factor enabling the big tech booms. Even though the U.S. economy was already in a great condition, in the 1970s, a group of neoliberal economists with ties to the University of Chicago began to argue that markets would do a better job if they regulated themselves. One of the biggest proponents of this claim was economist Milton Friedman. Another was jurist Robert Bork, who argued that "consumer harm" (high prices) was the only important criterion for an antitrust violation.

These men gained major influence with politicians who were already inclined to favor business interests. They spurred two government actions that would provide major benefits to tech companies: in 1978 Congress cut capital gains taxes from 49.5% to 28%, and in 1979 the U.S. Labor Department decided to let corporate pension funds invest in riskier asset classes.[10]

---

[a] Leland Stanford, founder of the university that has trained so many of today's technologists, was a tycoon in that industry.

At the start of the 1980s, Ronald Reagan and his administration gutted antitrust regulations,[11] as did leaders overseas such as Margaret Thatcher, Helmut Kohl, and Augusto Pinochet.[b]

Deregulation, and a lack of regulation in the first place, are major spokes in the Wheel.

Since the Reagan era, massive companies, from Walmart to Amazon, have been able to tower over the American economic landscape, and tech companies climbed to the top of the heap. By 2021, Google had managed to lock up over 90 percent of the worldwide search engine market, and Facebook held three quarters of the market in social media.[12] As of 2024, the most valuable companies in the U.S., ranked by market capitalization, were Microsoft,[c] Apple, Alphabet (Google), Amazon, NVIDIA, and Meta.[13]

Many journalists have written about how the Big Tech companies have attained near-monopoly status, yet when I look at the Comments sections I often see a shrug from readers. *Monopolies are only bad if they cause consumer prices to rise, so how can a company's domination of a market be a problem if they're selling their products at low cost, or giving away their services for free?*

This reflects a major misunderstanding of why monopolies can be harmful. Though Robert Bork and his neoliberal economist friends made consumer prices the only factor in deciding whether a monopolist needed to be reined in, the earlier antitrust regulators who broke apart goliaths like AT&T and Standard Oil were actu-

---

[b] Let's note that this was part of a significant long-term trend: though tech companies love to demonize government, they have often benefited greatly from its actions.

[c] Buoyed by its investments in A.I.

ally focused on whether the owners of such companies unfairly limited the ability of other businesspeople to compete.[d]

Libertarians argue in favor of letting companies operate without restriction, saying that this is the best way to encourage a thriving economy, but history has often shown the opposite. This marks one of the great ironies of a successful monopoly, whether it's a phone company or a social media app: while it might seem like the apotheosis of the capitalist system, it may actually work against the supposed tenets of that system. Says Rushkoff, "We have to remember this part of the program because it's so counterintuitive: the core code of the corporate charter is to repress exchange, competition, and innovation. It was intended to extinguish the free market."[14]

The new monopolists have learned that if they can provide consumers with services for cheap or free, they can get away with many of the offenses of the earlier robber barons. For one thing, they often crush smaller businesses. As Tim Wu points out, this is not a new trend in communications industries. The early Bell Telephone Company, for example, employed its profits as a war chest it could use to undercut the rates of many small local phone companies, "a tactic known as *predatory pricing*."[15] We've seen this more recently with Amazon, which sold books at a loss for years in order to undercut its competition, and with Uber, which started out offering bargain rates so it could undermine the taxi business.

The Big Tech companies also love to gobble up their rivals.

---

[d] "Monopolization" would seem to mean gaining total control over a market, but the Federal Trade Commission defines it differently, asking if a company has a leading position that was gained through improper anticompetitive conduct, "exclusionary or predatory acts"—"that is, something other than merely having a better product, superior management or historic accident." (https://www.ftc.gov/advice-guidance/competition-guidance/guide-antitrust-laws/single-firm-conduct/monopolization-defined)

Katy Cook notes that "Between 2007 and 2019, Google acquired over 270 companies, 171 of which were competitive acquisitions. In the same timeframe, Facebook acquired 92 companies, 46 which were competitors, almost all of which were purchased and then immediately shut down."[16] Wu calls this pattern the *Kronos* effect, in honor of the Greek titan who devoured his own children.[17]

Sometimes the ability to stifle competition doesn't require buying up rival companies; it just requires being the first to gain access to necessary resources. In today's race to rule the A.I. market, today's biggest companies already have a huge head start. For one thing, they've had the capital required to build the enormous data sets that the models rely on.[18] Operating a chatbot or other model also requires massive computing power. As tech journalist Cade Metz puts it, "Building a groundbreaking A.I. start-up is difficult without getting the support of 'the hyperscalers'"—companies such as Google, Microsoft, and Amazon—"which control the vast data centers capable of running A.I. systems. And that has put the industry's giants in the driver's seat—again—for what many expect to be the most important shift for the tech industry in decades."[19]

Cory Doctorow warns that when digital platforms grow so dominant, they can turn their areas of business "into 'kill zones' that investors will not fund new entrants for."[20]

Historically, in fear of competition for their products, monopolies have also tended to squelch innovation—sometimes even within their own ranks. Executives at AT&T's Bell Labs suppressed the development of their own magnetic audio recording system for the odd reason that they were worried that a fear of being recorded would inhibit people from using their telephones. Says Wu, "The recording machine is only one example of a technology that AT&T, out of such fears, would for years suppress or fail to market: fiber optics, mobile telephones, digital subscriber lines (DSL), facsimile machines, speakerphones—the list goes on and on."[21]

David Sarnoff, president of the powerful company RCA, which broadcast through AM radio, managed to suppress the superior FM technology for a whole decade.[22]

When a company grows so big that it has trounced its competitors, its leaders can grow arrogant and careless. In his book *Zucked: Waking Up to the Facebook Catastrophe*, Roger McNamee says that, "Zuck and Sheryl's failure to take action to address obvious flaws in the product and to protect their brand is at least suggestive of their monopoly power. They may not have been concerned about brand damage because they knew that users had no alternative."[23]

Even when we recognize the destructive power of monopolies, they're hard to escape. Cory Doctorow points out that, "The reason people who are aghast at Facebook's and Google's and Amazon's data-handling practices continue to use these services is that all their friends are on Facebook[e]; Google dominates search; and Amazon has put all the local merchants out of business."[24]

Maëlle Gavet notes that venture capitalists tend to "look for winner-take-all types of markets, because the chance for a large exit is higher. Particularly in these markets, the push for rapid growth can be extreme, sometimes known as 'blitz scaling'."[25] Investors might even aim for *hypergrowth*, which requires maintaining at least a 40% annual growth rate for more than a year.

The use of funding sources (like venture capital) which prioritize rapid, extreme growth over other

---

[e] Younger people may be on Instagram instead. (Which was bought up by Facebook.)

considerations, is a significant component in the Wheel.[f]

The growth imperative can result in business practices that would have seemed bizarre to the executives of traditional companies. As Gavet notes, "Desperate for growth, companies might spend more money to add customers than those customers yield in revenue, meaning they are just burning their investors' money. In effect, they spend $10 to acquire revenue worth $5. Or they might cut corners, or ignore regulators. Small, nagging problems are left to accumulate and fester. And in the frenzy to grow, they are less likely to focus on building empathetic, durable cultures, to treat their employees well, and to be good corporate citizens. All of these things damage the chances of long-term success."[26]

Eventually, the donkey can begin to drive the cart. One reason why growth has become so important to tech companies is because the business world loves metrics. Sandy Parakilas explains that, "In terms of design, they have intentionally built—and this is not just Facebook, this is Twitter, this is other similar companies, too—systems that are designed to enable them to grow users and collect data as fast as possible. [...] The reason it's built that way is because the stock market measures them based on revenue and user growth, so they literally take the metrics the stock market is going to measure them against and they say, 'Well, what are the features that will move these metrics the fastest?' They build those features."[27]

This reckless push is driving many of the problems we're seeing with today's tech.

When I spoke with former financial tech executive Robert Gardiner, he mentioned some problems he has witnessed:

---

[f]   There's nothing inherently wrong with a model that involves investors helping to incubate new companies, but problems arise if they push their startups in adverse directions for the sake of massive profits.

"You start out with a thing that is a very clever solution to a problem, and then that thing becomes so successful commercially that people want to scale up the business. *If only we could do ten x of what we're doing now, or a hundred or a thousand x.* There's a phenomenon that happens with scale, and you can see it played out in so many industries: with scale comes stupidity—a blurring. All of that focus and all of that knowledge that was about these instruments and how risky they were and how they worked—all of that knowledge that was embodied in people who were talking to other people gets diffused across a process, an assembly line.

I think that that so often plays out in technology, and it makes the prediction process very hard sometimes, because the scenarios you're trying to account for—it's ultimately a failure of imagination. Certain scenarios just won't happen until you get to a certain scale. If you're not thinking of scale, then you're not thinking hard enough about all the bad things that can happen."[28]

It's hard to overestimate the impacts of the growth imperative. "What is new here," Rushkoff says, "is that by applying our technological innovations to growth above all else, we have set in motion a powerfully destabilizing form of accelerated capitalism. We are worsening the disparity between rich and poor, punishing those who actually work for a living, losing human control over the capital markets, and letting shortsighted investors stifle long-term innovation."[29]

As he concludes, "Somehow, growth has become an end in itself—the engine of the economy—and human beings have come to be understood as impediments to its functioning. If only people and our idiosyncratic demands could be eliminated, business would be free to reduce costs, increase consumption, extract more value, and grow bigger."[30]

Antonio Garcia Martinez, who had a firsthand view of Facebook's relentless drive to attract users all over the world, was moved to quote environmental activist Edward Abbey, who wrote that, "Growth for the sake of growth is the ideology of the cancer cell."[31]

Roger McNamee has noted that, "Our culture has changed dramatically over the past fifty years, nowhere more so than in business. While management guru Peter F. Drucker once taught executives to balance the interests of five stakeholders—shareholders, employees, the communities where the employees lived, customers, and suppliers—businesses today focus almost exclusively on shareholder value. That allows corporations to lay off thousands, ship jobs overseas, and engage in business practices that would have been off-limits fifty years ago."[32]

Prioritizing the profits of executives and shareholders over the needs of other stakeholders (including workers, customers, community members, and the environment) is another common factor in the Wheel.

As we saw earlier with Google, big shareholders such as venture capitalists can pressure young founders to change a company's whole business model: to make advertising sales the ultimate goal, to exploit its users' data, to give up on a code of conduct that said "Do No Evil."

Sadly, such a single-minded focus on profits for shareholders has even been promoted by our legal system. "As corporate law is currently structured," notes Rushkoff, "CEOs and their boards of directors can be held liable if they fail to do everything in their power to maximize quarterly returns for public shareholders.

CEOs are not merely incentivized to pursue the short-term bottom line: they are legally obligated."[33,g]

On the other hand, this does not mean that tech leaders have surrendered all their power to shareholders and investors.

For example, Shoshana Zuboff has pointed out that, "One way that Google's founders institutionalized their freedom was through an unusual structure of corporate governance that gave them absolute control over their company. Page and Brin were the first to introduce a dual-class share structure to the tech sector with Google's 2004 public offering. The two would control the super-class 'B' voting stock, shares that each carried ten votes, as compared to the 'A' class of shares, which only carried one vote. [...] In the absence of standard checks and balances, the public was asked to simply 'trust' the founders."[34]

Let's add that kind of extreme top-down leadership to the Wheel.

Google is hardly the only company that has that kind of structure. Max Chafkin, author of a recent biography of Peter Thiel, has argued that the tech mogul's "most legendary bet—loaning $500,000 to a socially inept Harvard sophomore in exchange for 10 percent of a website called TheFacebook.com—is significant less for the orders-of-magnitude economic return he realized and more for the terms he embedded in the deal. Thiel ensured that Mark Zuckerberg would be the company's absolute dictator. No one, not even Facebook's board of directors, could ever overrule him. Similar maneuvers were adopted at many of Thiel's portfolio

---

g It's not only company leaders who are pushed toward a fast big killing. When engineers, coders, and other employees are given stock options, that may encourage them to fall in line and work toward the same unhealthy goal.

companies, including Stripe and SpaceX, and today, across the industry, it's more the norm than the exception."[35]

"Even the president of the United States has checks and balances," notes Tavis McGinn, a former Facebook pollster. "At Facebook, it's really this one person."[36]

Theoretically, the power of such company founders might counter the pressure from investors and shareholders, were it not for one simple fact: they're all aboard the same growth train.

Tech executives could make all sorts of changes that would render them better corporate (and world) citizens, but they often resist doing so—again, for that same basic reason. As John Naughton, a professor at the Open University in London, has put it, in a comment about social media companies that could apply to all sorts of firms, they "cannot solve the societal problems they have created—because, ultimately, doing so will hurt their revenues and growth'."[37]

"This," adds Naughton, "is the unpalatable truth they are all squirming to avoid. And in doing so they're really just confirming H.L. Mencken's observation about the impossibility of getting someone to understand a proposition if his income depends on not understanding it. It's not that the companies don't get it, just that they cannot afford to admit that they do'."

Tarnoff and Weigel's anonymous startup founder has explained how this typically plays out on the ground level—and in the heart of the Wheel:

> "An order comes down from on high: the board says to increase revenue. What's the best way the management team knows how to increase revenue? To increase time spent. So they issue the order, which gets percolated down the tree, and now everyone is

working on increasing time spent. This means making the product more addictive, more absorbing, more obtrusive. And it works: the user starts spending more time with the product. But every worker knows this is bad. Every engineer and designer knows this is awful. They're not happy making these features. But they can't argue with the data."[38]

Workers within tech companies have become increasingly aware of how such economic imperatives have impelled their bosses to promote all sorts of dicey practices. But what can they do with that knowledge? It's not easy to raise an internal alarm when you know, as one worker told the *Wall Street Journal*, that you're "standing directly between people and their bonuses."[39] It's even harder to speak out when you know that doing so might cost you your job.

Despite all of these obstacles, though, we may be seeing reasons for hope. There are a lot of ways in which concerned workers—and the rest of us in the general populace—might be able to push back. We'll spend the rest of the book considering them.

Part Three

SOLUTIONS

Much of the history of Silicon Valley has been a story of entrepreneurs rushing to release new technologies. Facebook provides a classic example. Here's Ruchi Sanghvi, the main programmer for the initial release of its News Feed feature in 2006, talking about how things worked in the company's early days: "With the push of a button you could push out code to the live site, because we truly believed in this philosophy of 'move fast and break things.' So you shouldn't have to wait to do it once a week, and you shouldn't have to wait to do it once a day. If your code was ready, you should be able to push it out live to user."[1] Adds Katie Geminder, the company's first product manager, "It really was just shove it out there and see what happens."[2]

We saw what happened.

But what if tech companies—and governments, and other interested parties—devoted more effort to predicting the consequences that might ensue from their decisions? How many of those problems could have been avoided, or at least reduced? More importantly, how might we anticipate and prevent future problems related to new technologies such as quantum computing, genetic engineering, and A.I.?

There are two main requirements here.

The first is a willingness to take the time and effort to really consider all possible consequences (not just positive ones).

The second is better predictive techniques.

There's good news on both of those fronts.

First, the recent techlash seems to have sobered a lot of workers in tech recently, and convinced them that it would be a good idea to try harder to foresee the consequences of their decisions.

For example, though the inventors of the horseless carriage probably didn't have the slightest inkling that their creation would one day monopolize vast amounts of land for roads and parking spaces, or dramatically reshape geopolitics for more than a century to come, analysts have recently done a lot more to foresee possible effects of its new descendant, the driverless vehicle. They have predicted, for example, that it might enable a major increase in highway safety, a reduction in traffic jams, and greater fuel efficiency. On the other hand, they have also envisioned that it could present new opportunities for terrorism (via hacked onboard computers), erosions of privacy (due to the vehicles' ability to collect data), the end of driving schools and of the auto insurance industry, and losses of municipal revenue from traffic tickets.[3,a] The results would be mixed, but the important thing is that we're considering such possibilities now, before the technology becomes widespread.

Here's another area where we seem to be making a stronger effort of prediction: while only a few savvy tech critics like

---

[a] It might also disrupt the lucrative business of drive-time radio, since all of the humans in the cars would be free to watch TV or surf the Web instead.

Douglas Rushkoff seemed to have been thinking deeply about what might happen with technologies like social media when they were first being created, today it seems that everyone and their cousin is vying to foresee what might happen with A.I.

The second piece of good news is that we now have a quiver packed with useful techniques that might help us do so. We'll spend the next three chapters considering them.

Chapter 12

Science Fiction:
Sounds Of Thunder

As I noted in the Introduction, complex causal sequences can be hard to predict, especially if they extend far into the future. Historian James Burke's book *Connections* is full of examples. For one, "The invention of the barometer and the discovery of air pressure suddenly multiplied the number of possible routes that the path of innovation could take. The event led to investigation of the behavior of gases, the discovery of oxygen and the development of respiratory medicine. It also led to hot-air balloons and the jet engine. It led, through the interest in how light rays—and later other types of rays—behaved in gases, to the cathode ray tube and to radar."[1,a]

Such compound chains, in which one initial action sets off a cascade of unexpected consequences, are often referred to as examples of the *Butterfly Effect*, a term derived from a 1952 Ray Bradbury short story called "A Sound of Thunder." The tale

---

[a] I find this information fascinating, but have to note how Burke couches it in deterministic terms, with technologies "leading" to other technologies, without mentioning human decisions and actions.

features a time-traveling big-game hunter who visits the age of the dinosaurs, accidentally tromps on a butterfly, then returns to his own era to find that he has initiated a long, subtle chain of effects that changed the result of the last presidential election.

There's no way that hunter could have foreseen the very distant results of his literal misstep, but science fiction authors have been coming up with successful predictions for centuries. In fact, they've foreseen a broad range of inventions, including robots, television, credit cards, teleconferencing, drones, antidepressants, earbuds, video surveillance, bionic limbs, and the internet.[2,b]

These authors not only managed to predict future technologies; sometimes, they even spurred inventors to *develop* them. Countless tech creators were motivated by childhood viewings of *Star Trek*, for example, and its spaceship's immersive playroom called the *holodeck* inspired developers of the metaverse.[3,c] Neal Stephenson coined the latter term in his 1992 novel *Snow Crash*, and his concept then influenced digital entrepreneur Philip Rosedale when he was developing the internet virtual world *Second Life.*[4]

On the other hand, sometimes, tech creators' takeaways from their reading seem oddly selective. Mark Zuckerberg proudly borrowed the term *metaverse* for what he was so eager to build, yet Stephenson's original conception was far from utopian. In *Snow Crash*, many public functions such as policing and local govern-

---

[b] Oddly, most failed to predict the ubiquity of cellphones. William Gibson, who coined the term *cyberspace* in his 1984 novel *Neuromancer*, notoriously had an artificial superintelligence in that book—which is set around 2035—try to contact the protagonist by ringing a row of pay phones.

[c] Fiction inspires fact—I'm reminded of how real Mafia dons began emulating the speech and behavior of the characters in *The Godfather*. (See the Mar. 9, 2022 *New York Times* article by Michael Wilson titled "With 'The Godfather,' Art Imitated Mafia Life. And Vice-Versa." The subhead reads "Federal wiretaps and Mob insiders suggest that many real-life wiseguys turned to the 1972 film for inspiration, validation, and cues on how to speak and act and dress.")

ment have been taken over by branded corporate entities, and the moment humans enter its virtual reality, they're beset by "animercials" that "swoop down on [them] from all directions, like buzzards on fresh roadkill."[5]

Zuckerberg missed the point (or perhaps he liked the idea), but in retrospect, this seems like a great example of a writer predicting a likely outcome. Just a year after Stephenson's novel was published, advertisers hit the internet like a swarm of buzzards, and within a decade, search engines, social media platforms, and other websites began using ad sales as their main source of income.

Like Stephenson, many sci-fi authors, such as Jules Verne, H.G. Wells, Aldous Huxley, and Phillip K. Dick, can be considered visionaries for what they got right about the future. For our purposes, their work is particularly helpful because they tend to foreground how human beings make choices to build technologies, and to imagine how people (and other beings) might be affected by those choices.[d] By doing so, they can help us become more focused on possible risks. Referring to a TV movie that had a major impact in 1983, Steven Johnson told me that, "*The Day After* was a big deal. I remember watching it when I was fourteen or something. It was a powerful mobilizing force, forcing you to confront the reality of what nuclear war would be like."[6]

Kim Stanley Robinson provides another great example of this galvanizing effect in his recent sci-fi novel *The Ministry for the*

---

[d]   When I teach thesis writing to design students, I suggest that they begin their papers by telling a story about a person experiencing a problem, in hope that this will encourage them to think of design and engineering in terms of human needs and impacts. We have all had the misfortune of encountering design that doesn't take the user's actual experience into account. I like to share an excerpt from a book by noted designer Don Norman called *The Design of Everyday Things*. It's from a chapter titled "The Psychopathology of Everyday Things," and it describes how something as simple and basic as a door can be designed so poorly that it frustrates users. (The book was published by Basic Books in 2002, and the passage is on pages 2-4.)

Future—he uses storytelling to get the reader to think deeply about global warming, a subject many people prefer to ignore. The book starts with a graphic description of a heatwave in India that kills millions of people. Envisioning such disasters might engender a sense of hopelessness, yet Robinson manages to fire up the reader by foreseeing a whole bunch of things we might do to solve the problem.

Fictional scenarios can help us get around a basic limitation of human psychology. In his novel, Robinson describes "a universal cognitive disability, in that people had a very hard time imagining that catastrophe could happen to them, until it did."[7] The recent worldwide spate of wildfires and extreme weather is obviously bad news, but there may be a silver lining: the immediate presence of such disasters may finally rouse more people to become concerned about global warming—and willing to do something about it.

Here's another example of why we might miss seeing what's ahead of us. For a few worried people, the Covid-19 pandemic did not come as a surprise. A variety of experts had predicted it,[e] and in 2019 the U.S. government actually conducted an exercise called Crimson Contagion, featuring many different agencies, that was designed to test its capacity to respond to a potential respiratory virus-based pandemic originating in China. The conclusion was that we were not adequately prepared.[8] We can assign some of the blame to specific people who ignored or downplayed the

---

[e] In 2018, Bill Gates warned that a pandemic might happen within the next decade, and others who predicted the contagion included environmental scientist Vaclav Smil, infectious disease expert Michael Osterholm, virologist Robert G. Webster, and Dr. Luciana Borio, director of the White House National Security Council's team for medical and biodefense preparedness. (https://www.businessin sider.com/people-who-seemingly-predicted-the-coronavirus-pandemic-2020-3? op=1#dr-luciana-borio-of-the-former-white-house-national-security-council-nsc- team-responsible-for-pandemics-has-previously-warned-of-a-pandemic-flu-threat- 7)

risk,[f] but we can also point to another cognitive bias: humans have a major tendency toward short-term thinking. Despite dire warnings, a global epidemic struck many people as just a hazy future possibility. Before Covid emerged, governments around the world were facing all sorts of immediate crises, and even if most of them were far less serious, they were *present* issues, and seemed more pressing.

When it comes to technology, the short-term thinking is often extreme. Steven Johnson cites "the 'minimally viable product' concept that is fashionable in the tech sector today: Don't try to ship the perfect product; ship the simplest product that might possibly be useful to your customer, and then refine and improve it once it's out in the market."[9] We're seeing a version of this right now with some companies rushing to publicly release A.I. chatbots, and saying that they'll put in full safeguards *after* they see what might go wrong.

This brings up an important point: the crux is not just *how* to make better predictions, but also *when*. **The time to try to anticipate possible bad consequences of a new technology is *before you build it.***

Many technologies undergo a phenomenon known as *lock-in*. In the digital world, when one program or system becomes popular, coders begin to write software to supplement it, and developers build apps around it, and before you know it, it becomes extremely difficult to change the fundamental nature of the design.

In general, of course, companies *want* their new creations to lock in, since that can help them build monopolies—they can effec-

---

f   As when Donald Trump's national security advisor John Bolton disbanded the White House's National Security Council Directorate for Global Health Security and Biodefense in 2017. Some of its staffers moved to other NSC posts, but see "Gross Misjudgment: Experts Say Trump's Decision to Disband Pandemic Team Hindered Coronavirus Response," by Deirdre Shesgreen, March 18, 2020 in *USA Today*.

tively lock other companies *out*. Johnson points out that "when a new service like Uber starts to take off, there's a strong incentive for the marketplace to consolidate around a single leader. The fact that more passengers are starting to use the Uber app attracts more drivers to the service, which in turn attracts more passengers. People have their credit cards stored with Uber; they have the app installed already; there are far more Uber drivers on the road. And so the switching costs of trying out some other rival service eventually become prohibitive, even if the chief executive seems to be a jerk or if consumers would, in the abstract, prefer a competitive marketplace with a dozen Ubers."[10,g]

Lock-in doesn't just take place with certain programs or apps; it can set in with whole technologies. Ursula Franklin put it this way: "The early phase of technology often occurs in a take-it-or-leave-it atmosphere. Users are involved and have a feeling of control that gives them the impression that they are entirely free to accept or reject a particular technology and its products. But when a technology, together with the supporting infrastructures, becomes institutionalized, users often become captive supporters of both the technology and the infrastructures. (At this point the technology itself may stagnate, improvements may become cosmetic or marginal, and competition becomes ritualized.) In the case of the automobile, the railways are gone—the choice of taking the car or leaving it at home no longer exists."[11]

In fact, the automobile makes a perfect example of a tech-

---

[g] Unfortunately, some technologies that become locked-in are actually inferior, as when the VHS tape format beat out Betamax, or the mp3 format triumphed over a number of other technologies that produced better sound. Another example is the unwieldy QWERTY keyboard layout which I'm using to type this sentence. It was designed in 1873 to prevent mechanical typewriter typebars from jamming together, but it's still with us in our modern computers, even though they no longer have those typebars, and other layouts would enable us to type with greater ease and speed.

nology that was pushed to lock in. At certain inflection points, when we could have gone in the eco-friendlier direction of making public transit the core of our transportation systems, the auto industry-funded "highway lobby" fought ferociously against that— and won.[12,h]

If we want to avoid locking in harmful technologies, we need to consider their impacts before that can happen. A former Google engineer put it this way: "After you've already built the prototype is not really the time to start thinking about the ethical ramifications."[13]

Again, science fiction can be helpful because it encourages envisioning those ramifications with a good deal of advance notice —sometimes before a technology is even feasible.

The art form can be valuable for another big reason. As we're about to see, a lot of professionals are now heavily invested in real-world forecasting, but they tend to only operate in a limited range of fields, in hope of finding practical benefits. What we really need, when it comes to technology, is to devote more energy to predicting possible *negative* impacts, especially psychological, social, and political ones—and that's what science fiction often does.

Beyond the realm of fiction, we have a number of other excellent practices that can help.

Let's take a look at some, saving two of the most intriguing ones for last.

---

[h] Big Tech's current push for driverless cars continues to send us in this direction, away from shared mass transit.

Chapter 13
A Predictor's Toolbox

"As the 9/11 Commission wrote: 'Imagination is not a gift usually associated with bureaucracies ... It is therefore crucial to find a way of routinizing, even bureaucratizing, the exercise of imagination.' Only once we have done that will we be able to reconcile the critical with the chronic, operations with strategy, the present with the future."

—Professional forecaster J. Peter Scoblic[1]

Do you spend any time in group meetings?

If so, then you have likely observed how overconfident or bossy participants can hog the discussion, or how you might feel pressure to overcome your personal objections and agree with the group. Now imagine how that might play out if you're in a roomful of people trying to make an important prediction.

In 1944, the U.S. Army Air Corps addressed these problems when it ordered up a report on how to foresee the effects of new technologies on warfare. Its researchers developed a technique called the ***Delphi method*** (still in use today), which involves asking a group of experts to make a forecast, providing them with a summary of the results, then asking them to revise their answers—and repeat the process, if necessary.[2] A key feature is that the participants offer their opinions separately and anonymously, to avoid the peer pressure that can arise in groups. (Often, they don't even know who the other contributors are.)

It's not surprising that military officials are invested in forecasting; for them, guessing right may be a matter of life and death. On a less lethal scale, businesspeople try to anticipate trends to help them figure out where future profit centers might lie, and campaign managers strive to predict election results. In the past seventy years or so, forecasting has become a multibillion-dollar industry, with hordes of consultants vying to provide advice about what's coming down the pike.

In the 1950s and 60s, military strategist Herman Kahn at the RAND Corporation and French futurist Gaston Berger at the Centre d'Etudes Prospectives were among those who developed the technique of ***scenario planning***. This involves imagining ways in which a situation might play out. The goal is to come up with a wide range of possible stories, rather than trying to settle on the most likely one. As Steven Johnson puts it in his book *Farsighted: How We Make the Decisions That Matter the Most*, "The three-part structure turns out to be a common refrain in scenario planning: you build one model where things get better, one where they get worse, and one where they get weird."[3]

Sometimes forecasters may be blinkered because they just start with existing facts and extrapolate from them. Scenario planning is designed to encourage decision makers to be more open-minded—to consider uncertain factors and the possibility of surprises. In the 1970s, Pierre Wack, an executive at Royal Dutch Shell, adapted the technique for business purposes, and his team managed to foretell what seemed like an unlikely event at the time

—an OPEC oil embargo—and thus enable the company to evade the financially brutal consequences when it was put into effect.[4]

"In a turbulent business environment," Wack wrote in 1985, "there is more to see than managers normally perceive. Highly relevant information goes unnoticed because, being locked into one way of looking, managers fail to see its significance."[5,a] He cited a major example from geopolitical history: "After agreeing to the nonaggression pact with Hitler in 1939, Stalin was so convinced the Germans would not attack as early as 1941—and certainly not without an ultimatum—that he ignored 84 warnings to the contrary."

Researchers in the military and business sectors have developed a bunch of other predictive techniques. These include ***simulation exercises***, sometimes in the form of ***war games***, and ***red teaming***, in which a group inside an organization play-acts the part of an aggressive competitor or adversary, to show how outside actors might affect possible outcomes. For example, cyber experts can pretend to be criminal "black hats" so they can discover how hackers might breach a company's defenses.[b]

Johnson chronicles how the Obama White House and the U.S. military and intelligence services devoted an incredible amount of time and effort in 2010 and 2011 to trying to decide if

---

[a] Wack became a bit disillusioned when he saw that corporate managers, even those at Shell who had seen great benefits from scenario planning, soon began to revert to their usual narrow ways of looking at the world. It was not enough to pay lip service to the technique by conducting a few brief workshops, as companies tended to do. (https://www.strategy-business.com/article/8220)

[b] The idea of red-teaming can be traced back at least as far as the Catholic Church's centuries-long practice of using "devil's advocates" to make the argument for the *con* side in hearings regarding the canonization of saints. (https://www.britannica.com/topic/devils-advocate)

Osama Bin Laden was really holed up in a compound in Pakistan; to figuring out the best ways to capture or kill him; and to predicting the political consequences of such an operation. Chastened by the shocking absence of such forethought on the part of their predecessors, who had rushed into the invasions of Afghanistan and Iraq, they were extremely cautious, relying on a slew of predictive methods that included scenario planning and red teaming, and they even went so far as to build a full-scale model of the compound in North Carolina, where they could game out what might happen if they used various tactics for an assault.

Notably, for our purposes in thinking about technology, such an approach might be especially helpful because it was explicitly designed to work against overly optimistic predictions. As Johnson puts it, "What this team had was different: a deliberation process that forced them to investigate all things they didn't like about the evidence and imagine all the ways their course of action could go terribly wrong. That process mattered every bit as much as the actual execution of the raid."[6]

As we learned afterward, of course, all that predictive work was worth the effort. This should lead us to ask, *If we can do so much to predict a military outcome, why can't we do more to predict technological ones?*

In 2023, the Def Con hacking conference in Las Vegas hosted a Generative Red Team Challenge, in which thousands of participants competed to "attack" generative A.I. models in order to find safety and security problems. A number of companies were involved, and the White House helped promote the event.[7] This seemed like a good exercise, and one that could be applied to other technological developments, but I think the timing was off. The teams in this challenge were working on models that had already been publicly released. Ideally, this kind of work should be done by companies *before they release their products*, and also before they *develop* them.

I realize this may sound seriously naive. It's clear why a defense establishment would devote such effort to ensuring that the success of a military operation, or why hackers would compete to win prizes and glory, but why would tech executives voluntarily invest time and labor in trying to anticipate the unintended impacts of their innovations?

For one thing, it might be a lot cheaper to use prediction teams in advance of releasing technologies, rather than dealing with dire consequences after the fact. Look at Mark Zuckerberg, who might have foreseen that there could be a great many problems inherent in his metaverse idea *before* he pumped billions of dollars into developing the technology, with calamitous effects on the company's stock.[c] Or consider how Elon Musk's 2022 takeover of Twitter might have proved less disastrous if he had been willing to let foresight teams warn him of how his changes might play out.[d] The benefits of better prediction could be reputational as well as financial—no one wants to be seen as the bad guy.

Of course, if they continue to turn a profit, some corporate executives won't care about bad consequences. (Consider the oil company honchos who covered up evidence that they were contributing to global warming,[8] and the cigarette company execs who covered up their products' ties to cancer.[9]) That's why a lot of the work of prediction ought to be taken on by outside entities such as governmental agencies, which can work to foresee trouble, impose regulations, and rein in companies before they can plunge ahead with potentially harmful actions.

---

[c]   Zuckerberg's massive, iffy investment in the metaverse helped drive the biggest historical one-day loss for a U.S. company: $230 billion in market value wiped out on Feb. 2, 2022. (https://www.theguardian.com/technology/2022/feb/03/facebook-stock-shares-meta-mark-zuckerberg)

[d]   On January 1, 2023, CNN reported that "Elon Musk Has Lost a Bigger Fortune Than Anyone in History." (https://us.cnn.com/2023/01/02/investing/elon-musk-wealth/index.html)

Since our governments are already charged with enforcing safety in such areas as aviation, automobiles, food, and pharmaceuticals, why can't they be empowered to do so for tech?

Aside from the techniques we've already looked at, some forecasters specialize in the field of ***threatcasting***, which focuses on what we can do now to avoid specific future dangers. ***Backcasters***, on the other hand, start with the question, "If we want to achieve a certain positive goal, what actions should we take to reach it?" ***Crowdsourcing*** and ***prediction markets*** aggregate a group's guesses about a possible outcome, such as whether a politician will win an election, or a stock will rise or fall.

Again, most of these techniques tend to be used by people in business, the military, or government. A notable exception was spurred by the environmental movements of the 60s and 70s, when activists were able to get legislation passed that mandated urban planning design and land use reviews before developers built major projects. These efforts often involve the use of *charrettes*, a collaborative method involving a series of small-group meetings that feature a wide range of stakeholders, including urban planners, architects, developers, government officials, residents of the area, and protectors of the environment. This marks a rare case in which the goal of professional prediction is something other than just increasing profit or winning conflicts, because regular citizens (or endangered plants and animals) might be seen as the beneficiaries.

A specialization that might prove helpful for our purposes is ***technology assessment***, which health technology professor David Banta has defined as "a form of policy research that examines short- and long-term consequences (for example, societal, economic, ethical, legal) of the application of technology." Banta notes that the term "came into use in the 1960s, especially in the United States, focusing on such issues as the implications of supersonic transport, pollution of the environment, and ethics of genetic screening."[10]

That sounds great, right?

Langdon Winner expressed a skeptical view, however, saying that "an unfortunate shortcoming of technology assessment is that it tends to see technological change as a 'cause' and everything that follows as an 'effect' or 'impact'."[11] He took exception to this way of looking at things, saying that "the attitude in which the predictions are offered usually suggests that the 'impacts' are going to happen in any case. Assertions of the sort 'Computerization will bring about a revolution in the way we educate our children' carry the strong implication that those who will experience the change are obliged to simply endure it. Humans must adapt. That is their destiny. There is no tampering with the source of change, and only minor modifications are possible at the point of impact."[12] He concluded with the rueful observation that then, "Social research boldly enters the scene to study the 'consequences' of the change. After the bulldozer has rolled over us, we can pick ourselves up and carefully measure the treadmarks."[13]

This points to one of the core themes of this book: **Prediction should not just be a passive exercise that allows citizens to foresee what is inevitably about to happen to us, but part of an active operation in which we can decide what future we really want**.

As promised, here are two techniques that offer special promise:

First, if you have worked in a bureaucracy, here's another situation that may seem sadly familiar: what if the organization you work for is plowing ahead with some new project and you can foresee bad unintended consequences, but the leader is an autocrat and you're afraid to speak out? (Who wants to get branded as a naysayer, a drag on the company's momentum?)

Wheel factor: a company culture that discourages employees from voicing concerns about the future impacts of a project.

Back in the late 1990s, a cognitive psychologist named Gary Klein developed a clever technique to avoid this all-too-common problem. He suggested that when company leaders propose a project, they should bring employees together for what he called a ***premortem***.

I'm sure you're familiar with the procedure called a *postmortem*; it means *after death*, and coroners use it to find out what caused someone to expire. We might also conduct one of these analyses if things have gone wrong in non-medical settings, as NASA did after the explosion of the space shuttle *Challenger*. But when we perform a technological postmortem, a plan has already been carried out; it's too late to change anything.

A *pre*mortem, on the other hand, takes place before a project has even been given the green light.

Klein asked participants to imagine that a plan had already been implemented, that it was now a year in the future, and they could look back and see that the outcome was a colossal failure. Their challenge was to come up with reasons for what might have gone wrong. Using this technique, no one gets shamed for sounding negative—in fact, they're rewarded for imagining unintended consequences.

When I spoke to Klein, now the head of a corporate training company called ShadowBox, he recalled an early success with the technique, when he consulted for an Air Force team that was planning a new combat system:

"Everybody was gung ho. We had the kickoff meeting; everybody was ready to go. Then we do the premortem and we say, 'This has failed; write down why.' And [there was] this young second lieutenant who had not said a word—this was a meeting that lasted two days. We went around the room and it was his turn and I said, 'What do you have at the top of your list?' He said, 'This algorithmic approach we're talking about, it takes forty-eight hours to run on a Cray computer, and we're talking about putting it on a laptop to go out into Iraq, where we need responses in just an hour or two. It's just a disconnect.' And after he said that, everybody looked at him and realized he was right—if we had continued with our original path, it would've failed. And then somebody said, 'You know, I have a quick and dirty method that I sometimes use if I don't want to wait for this to run on a Cray computer, and maybe we can sort of scale that up and use that.' And we said, okay, maybe we're saved. And that worked, and it became an extremely successful program."[14]

As its inventor has written, "The premortem doesn't just help teams to identify potential problems early on. It also reduces the kind of damn-the-torpedoes attitude often assumed by people who are overinvested in a project."[15,e]

Having had decades of experience with the corporate world, Klein told me that "most organizations are overconfident. Because frankly, you wouldn't start a business if you *weren't*. Given the base rates of how businesses or projects fail, you'd look at [that] and say, no, this is too hard. So in order to get anywhere, you have to have some degree of overconfidence to get you started. But you

---

[e] NASA has a term for this kind of eagerness to move forward: *go fever*. We can see it in the agency's rush to launch the *Challenger*, despite having been warned that there might be problems with O-rings on the shuttle's rocket boosters. (https://www.uml.edu/engineering/research/engineering-solutions/roger-boisjoly-challenger.aspx)

don't want to have so much that you become oblivious, and that's what happens. Organizations simply become unaware of their vulnerability and the brittleness of their plan and how things can fall apart so easily."[16] He notes that he has seen too many companies that will build a product "and they'll say here it is—and they won't imagine where or how it can fail."[17]

A premortem can help change that mindset, and it shouldn't just be a tool for business. Just think about how our recent history might have turned out if that kind of analysis had been used in 2003, before the U.S. invaded Iraq.[f] Or if—prior to 2016—our big social media companies had asked their employees to predict unintended political consequences of their new algorithms. Some junior engineer at Facebook might have felt able to raise her hand in a meeting and say, "Hey, if we show our users just the news sources that they and their friends already read, that might result in filter bubbles, which could encourage extreme partisanship and screw up our political process."

The premortem can show not only how things might accidentally go astray with a project or product, but what could happen if someone *intentionally* messes with it. Klein recalls an instance when his old company needed a new alarm system, and a security company executive told him that they had designed one in which the last person out of the building would punch in a code so the system would be armed.

"I said, 'What if I'm in the building and it's five o'clock and it's just me and somebody else and I happen to be really angry at the other person, and the other person is in his office working away, oblivious to the time, and I walk out quietly, I go to this pad, I

---

[f] A skeptic might argue that the Bush administration would have invaded the country anyhow, for political reasons—but at least we might have been better prepared for the massive unintended consequences.

punch in the alarm system, and then I leave the premises? Now this poor guy"—Klein laughed, remembering that potential scenario—"as soon as he walks out of his office, bells are going to go off, the police are going to be called, he's going to be in huge trouble. Could that happen?' And this guy who developed it had never thought about that before. He never thought about how somebody could be stupid or malicious, or just where the system could be corrupted."[18]

The lesson for us is huge: history has shown that just about any technology that is created for designated users and a particular purpose will likely end up employed by non-designated users for unplanned purposes. In other words, it's not just a matter of anticipating how *you* want your new technology to be used, but how *others* might use it—and that should include people with nefarious intentions. Anticipating such problems should be baked into the design process.[g]

Klein acknowledges that corporate CEOs may resist his technique, or apply it half-heartedly.[h] "They love optimism," he says. "They love people ready to charge forward." He adds, though, that if they use the premortem, it actually "makes them feel more rugged"[19]—it enables them to enjoy a more justified confidence about the soundness of their decisions.

He argues that if the technique is done right, it can have a significant positive impact on how corporations operate. "I remember doing a premortem with one company and there was a

---

[g] Steven Johnson cites our political Constitution and the blockchain technology as designs built with the possibility of bad actors in mind. (https://www.nytimes.com/2018/01/16/magazine/beyond-the-bitcoin-bubble.html)

[h] He says that sometimes executives will "use it just as a critique, and they'll say at the end of a meeting, 'I guess we should do a premortem. Does anybody see any problems here?' And that violates the whole spirit of a premortem, and then they don't get much value because they've done it wrong." (From an interview with the author.)

mix of senior leaders, the CEO, and a bunch of junior people, and we did the premortem and I thought it worked well. And then after it was over, people looked at me and they said, 'We have never had a meeting where people were so candid. We've never seen it before.' The premortem doesn't just allow it, it demands it. And it also changes the culture, because now people have seen that they can say things that might be atypical and it's not the end of their career, and you have senior leaders realizing that the junior people may have something to contribute."[20]

Using the technique takes some effort, but it can really pay off. "In the end," Klein once concluded, "a premortem may be the best way to circumvent any need for a painful postmortem."[21]

Now, you might (quite reasonably) be saying to yourself, Gee, some of these techniques sound pretty complicated and time-consuming. If we want to predict what will happen in a certain arena, wouldn't it be better to just ask an expert?

Unfortunately, your expert might well be wrong. This brings us to our second intriguing predictive technique: ***superforecasting***.

Starting in 1984, psychologist Philip Tetlock went on a mission to find out how to improve political judgment. He solicited hundreds of experts from a variety of fields to make predictions about future events—and found that, on average, their results were only slightly more accurate than predictions made by random guessing. In fact, he became notorious for arguing that "the average expert was roughly as accurate as a dart-throwing chimpanzee."[22] What's more, his research data showed "an inverse correlation between fame and accuracy: the more famous an expert was, the less accurate he was."[23] (I'm sure you can think of some well-known pundits who are often wrong.)

Tetlock began to see a connection to a quote from the ancient Greek poet Archilocus, who wrote that, "The fox knows many things, but the hedgehog knows one big thing."[i]

He realized that pundits tend to be hedgehogs, and they rarely acknowledge their mistakes—they sometimes get by on simply *sounding* sure. "A confident *yes* or *no* is satisfying in a way that *maybe* never is, a fact that helps to explain why the media so often turn to hedgehogs who are sure they know what is coming no matter how bad their forecasting records may be. [...] For example, people trust more confident financial advisers over those who are less confident, even when their track records are identical."[24]

So how do we help people make more accurate predictions? That was the challenge Tetlock faced in 2011, when a government agency called IARPA, for Intelligence Advanced Research Projects Activity, asked him, along with four other researchers, to engage in a geopolitical forecasting tournament. Each of them would put together a team that they could organize in any way they wanted, but they had to make predictions for the same set of very specific questions, such as, "Will the euro fall below $1.20 in the next year?" and "Will the president of Tunisia flee to exile in the next six months?"[25]

Counterintuitively, instead of putting together a team of hedgehog experts in fields directly connected to the questions, Tetlock gathered a crew of amateurs with a much broader range of interests. Over the four years in which IARPA ran the tournaments, he came to see that his best predictors, who he dubbed "superforecasters," came at their challenges with a very open mindset. "Stepping outside ourselves and getting a different view of reality is a struggle," he observed. "But foxes are likelier to give it a try. Whether by virtue of temperament or habit or conscious

---

[i] Technologists are often very specialized hedgehogs.

effort, they tend to engage in the hard work of consulting other perspectives."[26]

His best forecasters were often "ordinary" citizens—a retired Department of Agriculture employee, an unemployed factory worker, an artist. Yet they were able to make accurate predictions about subjects they knew little or nothing about, even though they had no access to insider information. In fact, they beat the other tournament teams, often composed of experts, by as much as 70% —and they were as much as 30% more accurate than intelligence community analysts.[27]

One thing that made them special was their way of thinking about a question. Tetlock promoted a method suggested by renowned physicist Enrico Fermi that consists of breaking the initial problem into a series of smaller, more manageable ones. For example, to answer the poser "How many piano tuners are there in Chicago?" without being given any further information, Fermi suggested starting with four other basic questions: *How many pianos are in Chicago? How often might they be tuned each year? How long might it take to tune one?* and *How many hours a year might the average piano tuner work?* Even if we don't have the actual answers to any of those questions, coming up with estimates for them can lead to a better approximation of the big answer.[28]

Superforecasters also have an advantage in that—unlike experts—they don't feel a need to hold onto and promote certain preconceptions and beliefs. Tetlock noted that, "The Yale professor Dan Kahan has done much research showing that our judgments about risks—Does gun control make us safer or put us in danger?—are driven less by a careful weighing of evidence than by our identities, which is why people's views on gun control often correlate with their views on climate change, even though the two issues have no logical connection to each other."[29]

Tetlock's amateurs also had the advantage that they weren't subject to the cognitive biases that come with being a part of a professional group. Psychologist Irving Janis, who had been one of Tetlock's PhD advisers, observed that "members of any small cohesive group tend to maintain *esprit de corps* by unconsciously developing a number of shared illusions and related norms that interfere with critical thinking and reality testing."[30]

Tetlock warned that such groupthink can lead to a politicization of science, in which "the goal of forecasting is *not* to see what's coming. It is to advance the interests of the forecaster and the forecaster's tribe. Accurate forecasts may help do that sometimes, and when they do accuracy is welcome, but it is pushed aside if that's what the pursuit of power requires."[31]

As Steven Johnson told me, more accurate forecasters "don't have a master narrative. They're not like, 'I know how the world works.' And so they're not constantly confirming that narrative. They're open to other possibilities." He added that they also tend to be "eclectic in their interests. In a way, there's a diverse group of experts inside their brain." [32]

Unlike most professionals, who have a lot of ego invested in being right, Tetlock's superforecasters were a lot more willing to consider that they might be mistaken. He noted that, "Researchers have found that merely asking people to assume their initial judgment is wrong, to seriously consider why that might be, and then make another judgment, produces a second estimate which, when combined with the first, improves accuracy almost as much as getting a second estimate from another person."[33]

One reason why sloppy pundits get away with being wrong is that a substantial amount of time often elapses between the prediction and the result, so their prognostications get forgotten before they can be proved wrong. One thing that made Tetlock's superforecasters more accurate was that they spent considerable time going back and analyzing their forecasts in light of the actual results. Unlike many pundits, they were very interested to figure out why they had erred—and they used that information to improve their future guesswork.

Tetlock noted yet another thing that made his superforecasters more accurate: they didn't limit themselves to focusing on the specifics of a particular case. "[They] are in the habit of posing the outside-view questions: How often do things of this sort happen in situations of this sort?"[34] When pressed for an example, he said,

"Well, if you're at a wedding and someone has the bad taste to come up to you and ask how likely is that couple to stay married, you're going to look at them [and say] look how happy they are. [You're] taking the inside view. [You're] basing [your] prediction on the particulars of that situation. [...] As opposed to [your saying], let's see, what's the sociodemographic category into which this couple falls? What's the divorce rate for that sociodemographic category? It looks like they've got about a, *mm*, sixty-five percent chance of staying married in the next five years and a fifty-three percent chance in the next ten years. That's the outside view. You're identifying a comparison class. And that comparison class is the start of your forecasting process."[35]

Unlike professional expertise, which often requires special education, years of experience, and access to insider information, superforecasting makes use of skills that anyone can develop. Tetlock found that if people took a crash course in how to do it, on average, they could learn to improve their accuracy by ten percent.[36]

How long did that training take?

Just one hour.

~

We have a lot of useful knowledge about how to make better predictions. Let's add it up, so we can come up with a better forecasting game plan—and one that will be helpful at preventing bad consequences in the world of tech.

For one thing, if we don't want to get blindsided by some unforeseen consequence of a new technology, we can take the outside view touted by Tetlock, and look for problems that have already occurred with related technologies in the past. For example, instead of just saying, *What do I predict will happen with* x *new communications technology in the future?* we can ask, *What does history tell us about what has happened with other communications technologies in the past?* (See Tim Wu's Cycle.)

Sometimes a negative consequence comes not as some diffuse, long-term effect, but the brutal swift punch of a disaster, as in the nuclear accidents at Chernobyl and Fukushima. Wendell Wallach says that such calamities involving complex mechanical systems tend to happen for four known reasons. He mentions incompetence or wrongdoing by managers or workers, including "the unwillingness of profit-seeking executives to implement costly safety systems": design flaws or vulnerabilities, including programming errors; dangerous interactions of poorly designed components, which often combine with human error in deadly chains and cascades; and underestimating the possibility of failures that are extremely abnormal and unpredictable—what risk researcher Nassim Taleb calls *black swans*.[37]

But some events that people think of as black swans were not unpredictable all. A number of experts had foreseen an attack like 9/11, and a number also anticipated the Covid-19 pandemic. In both cases, the predictions were correct; what was lacking was the political will to follow through with effective prevention.

Seeing how things have gone wrong in the past can help us avoid similar problems in the future, and that's one reason why we might want to include older people in our prediction teams: they have more direct experience of history. Steven Johnson has mentioned "the utility of youth in the context of innovation being that you have very little expectations about how things are supposed to happen," but he told me that, "it doesn't give you that long-term perspective—the sense of, 'Hey, you remember that big mistake we made thirty years ago when we rolled out that other technology? Like, couldn't we be making that mistake again?'" As with design charrettes, he recommended bringing in "people who represent a whole range of backgrounds and approaches and nationalities and fields of expertise," and added, "That's why I've tried to emphasize that generational diversity is incredibly important."[38]

The superforecasters were certainly a diverse bunch, but Tetlock observed that they "often tackle questions in a roughly similar way—one that any of us can follow: Unpack the question into components. Distinguish as sharply as you can between the known and unknown and leave no assumptions un-scrutinized. Adopt the outside view and put the problem into a comparative perspective that downplays its uniqueness and treats it as a special case of a wider class of phenomena. Then adopt the inside view that plays up the uniqueness of the problem. Also explore the similarities and differences between your views and those of others."[39]

He notes that after the Bay of Pigs debacle, President John F. Kennedy ordered a study that found that groupthink in the planning phase had been the main problem. The researchers suggested several solutions: "Bring in outsiders, suspend hierarchy, and keep the leader's views under wraps."[40] "A smart executive," Tetlock argued, "will not expect universal agreement, and will treat its appearance as a warning flag that groupthink has taken hold. An array of judgments is welcome proof that the people around the table are actually thinking for themselves and offering their unique perspectives."[41]

Johnson has observed that, "Our brains naturally project

outcomes that conform to how we think the world works. To avoid those pitfalls, we need to trick our minds into entertaining plot lines that might undermine our assumptions, not confirm them."[42] He recommends breaking decision-making into two parts, a *divergence* phase in which a wide range of possibilities gets considered, and then a *consensus* phase, in which we narrow down the options in hopes of agreeing on one path. He emphasizes that, in the initial phase, it's essential "to deliberately carve out a phase of the decision process within which entirely new alternatives are explored, to resist the easy gravitational pull toward the initial framing of the decision, particularly if it happens in the form of a 'whether or not' single alternative."[43]

He quotes psychologist Daniel Kahneman on the best way to run a group decision-making or predicting process: "You should always try to make these sources independent of each other. This rule is part of good police procedure. When there are multiple witnesses to an event, they are not allowed to discuss it before giving their testimony."[44]

"Before an issue is discussed," Kahneman suggested, "all members of the committee should be asked to write a brief summary of their position. This procedure makes good use of the value of diversity of knowledge and opinion in the group. The standard practice of open discussion gives too much weight to the opinions of those who speak early and assertively, causing others to line up behind them'."

Gary Klein told me that he recommends a technique called *gisting*, which was used by Tetlock and his research partner Barbara Mellers to make sure that all perspectives were being listened to. "Before I start trying to knock your ideas down, I've got to write maybe a paragraph or two coming up with the gist of what your arguments are. I've got to make sure I've understood you, rather than rushing in to try to critique you. And I think that's really valuable."[45,j]

Johnson observes that almost every good forecasting strategy he has examined "ultimately pursues the same objective: helping you to see the current situation from new perspectives, to push against the limits of bounded rationality, to make a list of things that would never occur to you."[46]

"The two things we will almost always benefit from," he concluded, "are time and a fresh perspective."[47]

Unfortunately, both elements tend to be in short supply in tech companies that rush products out in order to beat competitors, and have a homogenous, insular culture. That's why it was notable that in 2021, Meta touted the creation of a new Responsible Innovation team. Founded by V.P. of Product Design Margaret Stewart, the venture was designed "to help teams at Facebook proactively surface and address potential harms to society in all we build,"[48] and it featured a couple of dozen engineers, ethicists, anthropologists, civil rights specialists, and others, who collaborated with the company's product teams and consulted on their designs "as early as possible" in the development process. "In a given R.I. round table or workshop," Stewart added, "you might have a filmmaker, a philosopher, an artist, and an academic expert, alongside members of communities who could potentially be affected by what we are planning to build, all providing unique perspectives and wisdom to inform our approach."

"In the product design context," she explained, "this means thinking not just short- to mid-term, but investing time to forecast what longer term impacts might be. It means not just looking at the people who use the product as intended, but the people who may

---

[j] Thus technique could be valuable in other contexts too—such as romantic relationships!

misuse it to hurt others. It means considering if and how some people or communities may inadvertently have a negative experience with the products that we build."

This all sounds fantastic, right? It seems like a rare example of a Big Tech company wisely using the techniques we've been discussing in order to look ahead and prevent unintentional harm. On a company web page, Margaret Stewart wrote a post titled "Why I'm Optimistic About Facebook's Responsible Innovation Efforts."

Sadly, it seems that her enthusiasm was misplaced: just a year later, Meta was looking to do some cost cutting, and it shut the team down.[49]

Putting the job of prediction in the hands of one lone individual is a risky proposition. Unfortunately, as we have seen, the hubris driving many tech leaders renders them less likely to put together in-company prediction groups—and to listen to them if they do.

That's why governments should create vigorous programs designed to predict possible negative consequences of technologies, to help us avoid them before they can arise.

It's why universities and other institutions should do so.

It's why we citizens/users/customers should form our own prediction groups.

Ideally, it's why *all* of the stakeholders in the future of tech should be working harder to predict it.

We have plenty of tools.

Does the use of this technology arouse anxiety?
Does the use of this technology bring me joy?

Instead of just thinking of a technology as an end in itself, Sacasas pushes us to question what effects it will have on us as people—to ask, ultimately, whether it will further our human happiness, health, and moral development.

If a particular technology doesn't move us in those directions, then why should we create it?

To sum up this topic of prediction, I have asked if we can we make more thoughtful, well-rounded, and accurate predictions about the consequences of new technologies.

I believe we can. We just need the will to do so.

And then we need to ask not just *What are the technical possibilities of this new technology?* but *How might company leaders end up employing this technology in our real world, given the actual economic and political systems in which they operate?* And *How might it impact all eight billion humans, and the rest of life on Earth?*

We'll spend the rest of the book examining possible solutions to problems raised by our technologies, but first I'd like to recap the elements of the Wheel, in hope that it might provide a guide to danger signs that can help us avoid problems in the first place. To put it simply, if we want to predict the likelihood that a new technology will produce negative consequences, a good start would be finding out how many of these factors are present in its creation.

<u>ELEMENTS OF THE WHEEL</u>

<u>Cultural, Psychological, and Ideological Factors</u>

An overly homogenous culture where the technology is being created.

A very limited pool of people who get to decide what technologies are implemented.

A corporate culture that discourages employees from voicing concerns about future impacts.

Hubris and unchecked power, especially in company leaders.

A view of history that promotes the belief that technological progress necessarily equals human progress.

A tendency to put more faith in technology than in humans.

A shortage of emotional intelligence. (Especially a lack of concern for the effects of one's actions on others.)

Extreme techno-optimism, which ignores possible risks and bad uses.

Techno-solutionism. (The belief that almost all problems can be solved through technology.)

Beliefs in technological autonomy, determinism, and inevitability.

Randian ideology. (Anti-government, anti-regulation, anti-union, anti-social-supports, etc.) Also neoliberalism and techno-libertarianism.

Economic Factors

The use of funding sources—such as hyper-demanding venture capital—that prioritize extreme growth over considerations such as safety, the long-term viability of companies, and the well-being of workers and users.

Massive, rapid upscaling of networks, without commensurate oversight and safety measures.

A drive toward monopoly power.

Technological arms races against other companies.

A prioritization of the profits of executives and shareholders over the needs of other stakeholders, including workers, customers, community members, and the environment.

A lack of corporate investment in safety programs.

Corporate structures without checks and balances. (Especially dangerous when combined with founder hubris and immaturity).

An eagerness to use technology to replace human workers.

Demotion of users from the center of considerations to the periphery.

The offer of frictionless convenience and "free" services to mask exploitation of users.

Advertising as the primary means of making profit, if it involves aggressive invasions of user privacy, over-broad collection of user data, and extreme methods of boosting engagement.

Design Factors

Technologies that enable effortless, unvetted, unmonitored communication through vast networks.

Technologies designed to manipulate user psychology in unhealthy ways, including extreme attention-grabbing and the promotion of addictive use.

The use of uncontrolled computer algorithms to boost user engagement.

Providing too much ability for users to act anonymously.

Political Factors

A governmental inability to pass regulatory legislation (often spurred by intense industry lobbying).

Laws that remove responsibility and liability of tech companies.

A lack of legal and financial penalties for bad corporate behavior.

Technological arms races against other countries.

<u>Non-Predictive Factors</u>

A lack of interest and effort in predicting negative consequences.

A lack of organized mechanisms for prediction.

A lack of interest in asking questions about prospective technologies beyond technical feasibility and financial profit.

A lack of predictive effort before technologies are developed.

Of course, this list provides a negative way of looking at the situation. We can easily turn it around, though, and make it a positive checklist. To wit, creators of a technology will have a better chance of creating benefits for humanity and minimizing harmful consequences if:

They view technology as a means for serving human workers and human life, and the life of the planet in general (rather than viewing humans as a means for serving technology).

They balance optimism about new technologies with a realistic appraisal of risks and possible bad uses.

They encourage employees to consider possible harmful impacts of a technology, and enable them to voice concerns.

They make the effort to gain a well-balanced understanding of the history of technology.

They promote emotional intelligence in their leaders and workers, through hiring practices, training, company policies, etc.

They promote a diverse work culture in terms of age, gender, race, economic background, and disabilities.

They embrace ideologies that support the benefit of the many, rather than just the few.

They support vigorous in-house safety research and institute measures designed to promote safety.

They consult a broad range of people in deciding which technologies to implement, and how to implement them.

They appreciate that though some problems might be solved through technology, others might be best addressed through other means. (Such as social and political action.)

They seek out forms of financial backing that enable them to promote safety, the long-term viability of their companies, and the well-being of workers and users.

They work to build sustainable, fairly competitive companies, with just a reasonable level of profit.

They avoid massive, rapid upscaling, especially in networks, that supersedes the bounds of oversight and safety measures.

They seek out ways of earning revenue that do not rely on tricking users into believing that we're getting services without any cost; on manipulating user psychology and behavior in unhealthy ways (especially through the use of automated, uncontrolled algorithms); on monopolizing user attention; on encouraging addictive behaviors; and on unreasonable surveillance of users and extraction of their data.

They build corporate structures with checks and balances, so that the leaders don't have absolute, unquestioned power.

They serve the needs of a broad range of stakeholders. (Not just investors/shareholders.)

They focus on providing good services and products for users (rather viewing them as fodder for exploitation.)

They make considerable efforts to ask a broad range of questions about the possible effects of their technologies, including negative ones, and they create organized mechanisms for predicting them. They make these efforts *before* developing new technologies.

We'll also be better able to promote benefits and lower risks if:

Our government officials pass legislation to protect the rights and safety of consumers of technology, and the life of the planet.

Our government officials pass legislation that requires appropriate responsibility from companies for the technologies they develop, makes them liable for harms they cause, and imposes effective legal and financial penalties.

Our government officials create more powerful mechanisms for predicting the effects of new technologies.

Both tech creators and government officials make every possible effort to avoid starting and getting caught up in techno-political arms races.

Chapter 15
A Shared Fight

"Although effective contest will require determined individuals, the individual alone cannot shoulder the burden of justice, any more than an individual worker in the first years of the twentieth century could bear the burden of fighting for fair wages and working conditions. Those twentieth-century challenges required collective action, and so do our own."

—Shoshana Zuboff[1]

IN THE TV SHOW STAR TREK: THE NEXT GENERATION, a worlds-conquering cybernetic race called the Borg warns humans that, *"Resistance is futile."*[a] For those who would challenge new technologies, the response from Big Tech is often the same. It serves its leaders very well to have a populace that believes that

---

[a] The message actually appeared a number of earlier times in pop culture, from *Dr. Who* to *A Hitchhiker's Guide to the Galaxy* to an *X Men* comic to "Monsters of Moyen," a 1930 short story by one Arthur J. Burks.

there's nothing we can do to stop the march of technological "progress."

Throughout history, though, people *have* occasionally managed to contend against problematic new inventions. In this chapter I'll focus on some defiant predecessors and current resisters, and then—for the rest of the book—we'll examine a bunch of possible solutions to our technology-based woes.

We have seen how social scientists pointed out major inaccuracies in the traditional story of hunters and gatherers and the beginnings of agriculture and other technologies. They had other issues with it as well—and here we come to a third view of human history.

For one thing, the timeline is flawed. James C. Scott notes that, "It turns out that sedentism[b] long preceded evidence of plant and animal domestication and that both sedentism and domestication were in place at least four millennia before anything like agricultural villages appeared. Sedentism and the first appearance of towns were typically seen to be the effect of irrigation and of states. It turns out that both are, instead, usually the product of wetland abundance."[2] Many of the so-called "cradles of civilization" were naturally irrigated, from the Fertile Crescent in the Middle East to the Nile River Valley to North America's Ohio River Valley, and their inhabitants seem to have practiced both farming *and* hunting and gathering.

It's easy to imagine early proponents of the technologies of agriculture smugly proclaiming that, *You Can't Stop the Plow*, yet, like Jared Diamond,[3] David Graeber and David Wengrow point to a number of cultures around the world that opposed sedentism and farming for hundreds or even thousands of years. As Scott put it, "It is simply assumed that weary Homo Sapiens couldn't wait to

---

[b] The practice of living in one place for a long time.

end hundreds of millennia of mobility and seasonal movement. Yet there is massive evidence of determined resistance by mobile peoples everywhere to permanent settlement, even under relatively favorable circumstances. Pastoralists and hunting-and-gathering populations have fought against permanent settlement, associating it, often correctly, with disease and state control."[4]

In the traditional version of history we can also detect strong notes of *technological determinism*, the notion that the development of new technologies inevitably leads to certain particular results, like agriculture "giving rise" to monumental architecture, and the advent of city-states. But Graeber and Wengrow show that humans were building substantial towns with monumental architecture *before* the spread of agriculture, such as Gobekli Tepe in southeast Turkey, in 9,000 B.C.[5] They also note that, "Roughly 6,000 years stand between the appearance of the first farmers in the Middle East and the rise of what we are used to calling the first states; and in many parts of the world, farming never led to the emergence of anything remotely like those states."[6]

In another big revelation, Graeber and Wengrow demonstrate that agricultural settlement didn't necessarily lead to social inequality and the rise of kings and queens. Monarchs tended to leave behind statues and monuments and other grand self-tributes, to have themselves buried with fabulous riches, and even—in a number of cultures—to have their own families and retinues killed in order to provide them with company in the afterlife. The authors point to a number of prominent archaeological sites that appear free of such evidence of top-down rule, such as Minoan Crete, the Hopewell culture based in North America's Ohio River Valley, and Teotihuacan in Mexico, whose residents seemed headed toward monarchy, building grand monuments such as the Pyramids of the Sun and Moon—only to reverse course and concentrate on providing good housing for all of their citizens.[7]

The authors argue that "many of the first farming communities were relatively free of ranks and hierarchies. And far from setting class differences in stone, a surprising number of the world's earliest cities were organized on robustly egalitarian lines, with no need for authoritarian rulers, ambitious warrior-politicians, or even bossy administrators."[8] As for agriculture leading inevitably to the notion of private property, they point out that, "If anything, flood-retreat farming was practically oriented towards the collective holding of land, or at least flexible systems of field reallocation."[9]

"Our conventional vision of world history," write Graeber and Wengrow, "is a chequerboard of cities, empires and kingdoms; but in fact, for most of this period [the last 5,000 years] these were exceptional islands of political hierarchy, surrounded by much larger territories whose inhabitants, if visible at all to historians' eyes, are [...] people who systematically avoided fixed, overarching systems of authority."[10]

In short, the traditional story is one of people everywhere kowtowing to the "inevitable" consequences of a new technology. But a more accurate version shows that we have some choice in the matter: time and time again, all over the world, people have seen how a new technology might change their lives in ways that were not necessarily for the better, and they either flat-out refused it, or only accepted it in limited ways.

Moving forward in time, we can see other examples. As I've noted, many technologies have been developed for military purposes. Occasionally, weapons have been renounced on grounds that they seemed morally offensive. Neil Postman noted that, "The use of the crossbow was prohibited, under the threat of anathema, by Pope Innocent II in the early twelfth century. The weapon was judged to be 'hateful to God."[11] Lewis Mumford mentions a case in which a possible weapon was even rejected by its own inventor:

"Though Leonardo [da Vinci] wasted much of his valuable time in serving warlike princes and in devising ingenious military weapons, he was still sufficiently under the restraint of humane ideals to draw the line somewhere. He suppressed the invention of the submarine boat because he felt, as he explained in his notebook, it was too satanic to be placed in the hands of unregenerate men."[12]

Sometimes, resistance to technologies was ordered by the leaders of political systems. For example, although by the 15th century the Chinese had built a great fleet of ocean-going ships, they abandoned them for several hundred years because they felt that the technology brought in global influences that threatened their economy and cultural identity.[13]

Jumping ahead several centuries, we encounter a workers' movement in England that has since become synonymous with resistance to technology. In the world of tech these days, a preferred way to respond to anyone with doubts or concerns is to claim that they oppose innovation due to fear of technology, and to scornfully dismiss them as "Luddites."[c]

Those who wield the term that way misunderstand its history. As Neil Postman describes it, "between 1811 and 1816, there arose widespread support for workers who bitterly resented the new wage cuts, child labor, and elimination of laws and customs that once protected skilled workers. Their discontent was expressed through the destruction of machines, mostly in the garment and fabric industry; since then the term 'Luddite' has come to mean an almost childish and certainly naïve opposition to technology. But the historical Luddites were neither childish nor naïve. They were people trying desperately to preserve whatever

---

[c] The movement was supposedly inspired by one Ned Ludd, a young textile mill laborer who smashed a new knitting machine that had replaced many workers. It seems, though, that he may have been an invented, symbolic character, somewhat like Robin Hood. (https://www.smithsonianmag.com/history/what-the-luddites-really-fought-against-264412/)

rights, privileges, laws, and customs had given them justice in the older worldview."[14]

Calestous Juma, a Harvard professor of International Development, pointed out that, "People do not oppose technologies simply because they are new or because they are ignorant. They resist loss. The loss can be in the form of income, identity, worldview or power."[15] He added that "resistance to new technologies is intensified by the perception that its benefits are likely to accrue to a small section of society, while the risks could affect a wider section of society." I find it hard to imagine a more concise expression of our current situation *vis a vis* artificial intelligence, in which millions of people might lose their jobs, yet the technology could help a few fabulously wealthy tech moguls become even richer. (Investor Mark Cuban predicts that the world's first trillionaires will be entrepreneurs who have mastered A.I.[16])

The Swing Riots, which took place in England between 1830 and 1832, constituted an even larger revolt than the Luddite one. This time, the protesters were agricultural workers, and one of their complaints was the introduction of mechanical threshing machines. Since a single person operating one could replace ten doing the job by hand, the threat to workers was considerable.[17]

Speaking of farmers, I'd be remiss if I didn't mention the Amish, who are probably America's best-known living rejectors of technology. I've saved them for the end of the chapter, where we'll see that—as with the Luddites—their reasons for doing so have been largely misunderstood.

Back on our timeline, here's an instance of workers who got fed up with what Ursula Franklin called *control-based* technologies: in the 1930s, Ford Motor Company workers in a Brazilian rubber processing facility rebelled against the founder's extreme emphasis on monitoring them for efficiency. As they protested, they smashed the time clocks that were used to track their entrances and exits.[18] (I can't help but wonder if modern Amazon fulfillment center workers might someday follow suit, rising up against the managers who use surveillance devices to track their every move, and fire them for not completing enough assignments per hour, sometimes making them afraid to take "time off task" to use the bathrooms.[19])

Given some of the examples I've just mentioned, it would be easy to think that the rejection of technologies is just a cause for workers afraid for their jobs, but—as I mentioned earlier—a number of inventions have actually been refused by big companies, such as communications industries eager to suppress any new media that might threaten their hegemony.[d]

As I noted earlier, disasters can provide convincing evidence of a need for resistance, as the nuclear power plant meltdowns at Three Mile Island in 1979 and Chernobyl in 1986 showed. They boosted a worldwide nuclear freeze movement that greatly slowed the deployment of that power source.

Moral horror can also be a strong incentive. Inventors have recently developed weapons that would put the crossbow and submarine to shame, and protesters have had some success in restricting or blocking them, such as a 1997 international treaty that bans the use of antipersonnel mines, and a 1998 ban on military lasers that could blind soldiers and civilians. (Unfortunately, activists have not been able to secure a ban on autonomous lethal weapons.)

Other protesters have managed to achieve restrictions on genetic editing in humans, and performance-enhancing drugs and hormones for athletes, and a number of European countries have banned genetically modified organisms (GMOs). Sometimes, as

---

[d] Clearly, the rejection of a technology is not always a good thing; I'm just making the point that it *is* possible.

with human cloning or stem cell research, the objections have been religious in nature, based on the notion that technologists were usurping the role of God.[e]

In the digital age, users have sometimes balked after a company engineered a particularly obnoxious feature. Jaron Lanier cites Facebook's 2007 implementation of a hard-to-opt-out-of feature called Beacon. "When a Facebook user made a purchase anywhere on the internet," he explains, "the event was broadcast to all the so-called friends in that person's network. The motivation was to find a way to package peer pressure as a service that could be sold to advertisers. But it meant that, for example, there was no longer a way to buy a surprise birthday present. The commercial lives of Facebook users were no longer their own. The idea was instantly disastrous, and inspired a revolt. The MoveOn network, for instance, which is usually involved in electoral politics, activated its huge membership to complain loudly. Facebook made a quick retreat."[20]

"The Beacon episode cheered me," Lanier concluded, "and strengthened my sense that people are still able to steer the evolution of the net. It was one good piece of evidence against metahuman technological determinism. The net doesn't design itself. We design it."

Indeed we do, and sometimes a product fails due to poor design, or an unanticipated lack of consumer interest. Google Glass "smart" spectacles were supposed to be the Next Big Thing when they were released in 2013, but they failed for those reasons, plus privacy, safety, and security concerns. (Consumer advocates quickly realized that they could be used to surreptitiously record and broadcast conversations, photograph people without their

---

[e] I mention this not because I'm advocating for religious bans on technologies, but simply to show that resistance has come from many different quarters.

awareness, and even capture someone else's password as they typed it into an ATM.[21])

∽

We are reaching our current moment.

In a 2020 article titled "Hurting People at Scale: Facebook's Employees Reckon with the Social Network They've Built," reporter Ryan Mac showed that many of the company's workers were particularly upset about Mark Zuckerberg's failure to take action after Donald Trump posted "When the looting starts, the shooting starts" during protests over the murder of George Floyd. They had plenty of other corporate shortcomings to be troubled by, and things soon reached a point where an internal satisfaction metric showed that less than half of the employees agreed with the statement that "Facebook Is Making the World a Better Place."[22]

So what have tech workers done about the problems?

Some would argue, *not much*. Never one to mince words, Antonio Garcia Martinez said in 2018 that, "One thing that I think is annoying is the lack of moral courage in Silicon Valley. No one takes a stand on anything 'cause the opportunity costs can be so great given the winner-take-all nature of it, right? Missing out on being employee five at the next Uber is like a life-defining event and no one wants to fuck that up."[23]

I'm sure that's true in some cases, but there are also plenty of tech workers who got into the industry not just for the money, but because they loved solving engineering problems, and designing useful products. What's more—to quote the old Google motto— many of them care about not being evil.

Some would assert that even if employees want to do something about the problems their companies are causing, they're ill-positioned to bring about change. "We now know that many technology products are unsafe," Roger McNamee has written. "The aspects that make them unsafe are central to their economic value, which means that pressure for change must come from the outside."[24]

In fact, though, in the past decade, considerable pressure for change has come from the *inside* of tech companies, voiced by their own workers. In some instances, they've been able to band together and overturn objectionable corporate decisions, as when, in 2018, Google employees successfully pressured the company to abandon a search engine prototype called Dragonfly that would allow the Chinese government to censor search results and record users' searches.[25] They also convinced their top executives to not renew a contract that offered some of the company's A.I. capabilities for the U.S. Department of Defense's Project Maven, created to improve drone targeting in warfare.[26,f] Thousands signed a petition which said that, "We believe that Google should not be in the business of war."[27]

"Shortly after the internal uprising," notes Kate Crawford, "Google released its Artificial Intelligence Principles which included a section on 'AI applications we will not pursue.' These included making 'weapons or other technologies whose principal purpose or implementation is to cause or directly facilitate injury to people,"[g] as well as 'technologies that gather or use information for surveillance violating internationally accepted norms'."[28]

---

[f]   That didn't kill the project, however: it continued, thanks in part to some Silicon Valley startups whose boards include such moral paragons as Peter Thiel, former Google chief Eric Schmidt, and James Murdoch, son of Rupert. (https://www.forbes.com/sites/thomasbrewster/2021/09/08/project-maven-startups-backed-by-google-peter-thiel-eric-schmidt-and-james-murdoch-build-ai-and-facial-recognition-surveillance-for-the-defense-department/?sh=4879b1496ef2)

[g]   Sadly, this Silicon Valley trend of resisting the use of tech for war might be reversing: in 2023, *Washington Post* reporters found that "skepticism against defense work has faded for younger generations," with "battlefield technology driving a funding frenzy among tech investors." (https://www.washingtonpost.com/technology/2024/02/17/silicon-valley-military-tech-defense-contractors/)

Workers led other internal campaigns in 2018. When Microsoft employees protested the company's $19.4 million contract with the U.S. Immigration and Customs Enforcement agency due to inhumane treatment of migrants, they sent a letter to the CEO which said, "We are part of a growing movement, comprised of many across the industry who recognize the grave responsibility that those creating powerful technology have to ensure what they build is used for good, and not for harm."[29]

Also in 2018, a bumper year for tech protests, Amazon workers demanded that their company stop selling facial recognition software to law enforcement agencies, arguing that it could violate human rights.[30] (It has been shown to sometimes misidentify people of color, leading to false arrests, and there are concerns that it could be used to discriminate in hiring, in who gets loans, and in other areas.[31]) It took quite a fight, but they eventually got the company to put a moratorium on that use.[32] A number of municipalities across the U.S. also passed legislation banning the technology (although some of them later reversed course).[33]

In 2021, after worker protests, Facebook announced that—in the interest of maintaining users' privacy—it would end its use of facial recognition software and delete facial data on more than a billion people.[34,h] The company also announced that it was pausing the development of Instagram's new platform for children. Both instances provided rare and welcome examples of the fact that big companies *can* decide to reverse or restrict use of technologies, even after they have poured lots of money into developing them.

In a number of other instances, though, workers were disappointed to find that their internal activism couldn't force their

---

[h] Critics pointed out that the company had already done an enormous amount to normalize use of the technology by introducing a feature in 2010 that automatically tagged people in Facebook posts, even on photos that they didn't know that others had taken.

companies to really change course. In 2013, Tristan Harris, a design ethicist at Google, wrote an internal presentation called "A Call to Minimize Distraction and Respect Users' Attention." His slide show went viral among the employees, and he was given company support to continue his research, but he ultimately found that it had a limited impact. "At the end of the day," he said, "what are you going to do? Knock on YouTube's door and say, 'Hey, guys, reduce the amount of time people spend on YouTube. You're interrupting people's sleep and making them forget the rest of their life'? You can't do that, because that's their business model."[35]

Guillaume Chaslot recalled that, "As an engineer at Google, I would see something weird and propose a solution to management. But just noticing the problem was hurting the business model. So they would say, 'Okay, but is it really a problem?' They trust the structure. For instance, I saw this conspiracy theory that was spreading. It's really large—I think the algorithm may have gone crazy. But I was told, 'Don't worry—we have the best people working on it. It should be fine.' Then they conclude that people are just stupid. They don't want to believe that the problem might be due to the algorithm."[36]

Disappointed by the lack of response they received when they spoke up inside their workplaces, tech workers began to go public. When I spoke to journalist Noah Kulwin, he told me that talking to remorseful company insiders was "like Calvin's mom opening up the closet door and all of the toys falling out."[37,i]

When even public complaints didn't seem to be having enough impact, some workers, like Frances Haugen, became whistleblowers, releasing internal documents that revealed how executives were aware of problems but didn't act, or failed to do enough to bring about real change. And, like Haugen, some internal critics were so fed up that they quit, as Harris did in 2015

---

[i]   A reference to the comic strip *Calvin and Hobbes*.

when he left Google to form a nonprofit called Time Well Spent, later renamed the Center for Humane Technology.

Company workers may have little power to reverse their businesses' courses of action. That was not the case with Dong Nguyen, a 28-year-old independent video game designer who created a smartphone amusement called Flappy Bird that went viral in 2014, garnered 50 million downloads, and became the world's most popular free app. Due to ad income, he began earning tens of thousands of dollars a day, but he started receiving disheartening messages from unhappy users, including a woman who blamed him for "distracting the children of the world," a student who wrote that the game was "addicting like crack," and a worker who said that he'd lost his job due to playing the game too much.[38]

Nguyen went on Twitter and announced that he was removing the game from app stores.[39] In a subsequent interview, he said that, "Flappy Bird was designed to play in a few minutes when you are relaxed, but it happened to become an addictive product. I think it has become a problem. To solve that problem, it's best to take down Flappy Bird. It's gone forever."[40] Then he set about designing a new game that would bake ethical decision making into the design process: it would still be captivating, but less destructively so.

That's a rare tale in today's tech world, yet Adam Alter, author of a book called *Irresistible: The Rise of Addictive Technology and the Business of Keeping Us Hooked*, describes "a growing movement of ethical game design."[41] For example, some developers started to distance themselves from games that go on without conclusion, such as Facebook's popular Farmville. Alter quotes legendary game designer Bennett Foddy as saying that, "They're an abuse of a weakness in people's motivational structures—that they won't be able to stop. Instead [some designers] prefer to make games that engage you till you get to the end—and then it's over and you're free from it."

Through L.M. Sacasas, I learned of another principled rejection of technology. He offered me a link to a video presentation by Erhardt Graeff, an engineering professor, titled "The Responsibility to Not Design and the Need for Civic Professionalism."[42] Graeff was inspired by two groups of his students who were trying to develop tools that would benefit society. One came up with an internet data-collection program that they hoped would disrupt sex trafficking, and the other was planning a tech platform for a district attorney's office. The first group realized that their program might cause unintended harm to victims, and the second decided that the platform they were about to create didn't align with their ethical values. They resolved not to build their own projects.

Impressed by these examples of what he calls "design refusal," Graeff was moved to observe that, "Engineers ask, 'Can we build something?' But we also need to ask, '*Should* we build something?'" He noted that since the recent techlash, he had heard tech workers saying, "We should have a say in what we build, and we shouldn't be building [certain] things. I didn't go into tech to build surveillance tools, or algorithms for armed drones. There are arguments here that more people should be part of the conversation about what the tech industry should—and shouldn't—build."

We have just looked at historical instances of resistance, and recent examples of protest from within the tech community, but how can members of the public challenge today's problematic technologies?

One thing we can certainly do is to use tech more consciously on an individual basis. Instead of mindlessly jumping onto the Web if we're bored, or scrolling endlessly down Instagram or TikTok feeds, we can limit our consumption of digital media. Ironically, developers and engineers have recently created a new cottage industry of apps and features that can help us do that, from graying out our phone screens so we're less beguiled by the bright colors to providing timers that temporarily lock applications if we're spending too much time on them. (Even more ironically, Steve Jobs, Bill Gates, and a number of other prominent creators of tech, recognizing its dangers, limited their own children's use of it.[43])

Apple has included a *Screen Time* feature in its devices that enables users to track the amount of time they're spending on each app. (Of course, that's only if—and it's a *big* if—they choose to use it.) And I'm impressed by software engineer Josh Wardle, creator of the game Wordle, who explicitly designed his challenge so that it could only be played once every twenty-four hours. (I can testify that it's addictive, but spending three minutes a day on it is not going to make anyone lose their job.)

Speaking of a more conscious use of technology, I'm going to return to the Amish now, and to the inaccurate idea that they reject all of it because they want to remain in some naïve prelapsarian time.

"I would say that the Amish are not exactly resisting technology," Sacasas told me. "What they're attempting to do with varying degrees of success is to preserve the distinctiveness of their community. And I think this is actually an important point when we think about what we are trying to do with regards to technology. Resisting certainly, at some places, is the appropriate mode. But that's a reactionary way of framing it in some respects. What I would want to stress is that we not think about technology as a thing to resist, but rather that we highlight or focus on what it is that we desire—what are the goods that we desire for ourselves, for our communities?—and then think very carefully and critically about how a new tool, a new device, a new system would interact with these goods."[44][j]

Sacasas explains that this is what the Amish do. Rather than acting as knee-jerk rejectors of technologies, they evaluate them carefully. "They actually have individuals who we would sort of think of as early adopters whose role in the community is to adopt a new technology in order to explore its possibilities and then to make recommendations to the community as to whether this is a technology that can be adopted without affecting their larger community life; or whether it can be adopted but with severe restrictions or limitations; or whether it's one that's just best not integrated into the community at all. So it's not that they, as some people think, just sort of froze time in the 19th century and decided not to accept anything else, but rather that they decided they're going to negotiate the terms upon which they're going to allow technologies to enter their community."

That's why the Amish have rejected radios, TVs, and computers, which would introduce all sorts of foreign values into their culture, as well as cars, which might draw members away from its geographical centers, but are willing to use such helpful devices as gas grills, shop tools, batteries, electric fences, and copy machines.[45]

I should probably not have been so surprised to hear of their considered way of dealing with the issue, since I had already learned of the Amish practice called *rumspringa*. Instead of expecting their young people to just blindly accept their faith and its ways of living, when they reach the age of sixteen, their elders allow them a period in which they can act beyond their usual communal restrictions. Some try out typical American "worldly"

---

[j] To be clear, Sacasas is referring here to psychological and social benefits, not commercial goods.

behaviors such as dating, wearing non-traditional clothes, driving cars, going to movies, and using other forms of technology.[k] After having been given the option to experience and appraise other lifestyles, they can then make a conscious decision about whether they want to officially join the church.[l] This kind of thoughtfulness and willingness to test also informs the Amish approach to technology.

Sacasas explains that, rather than focusing on its negative aspects, this community has chosen to home in on what constitutes a positive way of living. "They have a high degree of buy-in, if you will, into a particular understanding of what counts as human flourishing. I've referred to them as being extremely tech savvy in that very long ago they recognized that the introduction of a new tool or a new technology doesn't just add something to their existing community—you're threatening to completely transform the social ecosystem. What they very early on seem to have understood—in a sense, what the Luddites understood with regards to *their* way of life—was that if you introduce certain technologies, it will necessarily change the dynamics of the community. And maybe for the worse."

I mention this not as some sort of unequivocal endorsement of Amish culture, nor to suggest that we all go back to driving horse-drawn buggies. For our purposes, what's intriguing is how these people have managed to engage with technology so consciously and proactively, rather than passively accepting whatever new inventions come along.

If the Amish can manage to do that, even in this tech-saturated age, that makes me wonder why we *all* can't take more control over which technologies we're willing to allow into our lives.

---

[k] The media has tended to hype the extreme behavior of a few, like having pre-marital sex or taking drugs, but most of the experimentation is considerably more tame.

[l] A great majority of them do.

As I've noted, we can make a number of choices on a personal level. There are all sorts of small things we can change in our relation to our devices and tech platforms. We can silence notifications (at least part of the time). We can refuse to read emails before going to bed. (I do that in order to make sure that I'm not going to read something that will disturb my sleep.) We can delete trivial apps that take up too much of our time. I have chosen not to enable email on my phone at all, so I can at least be free of it when I'm out and about. We can use automated email and voicemail responses to make it clear that we have set limits on our work hours and response times. We can be more selective about clicking on things online, to minimize the bombardment of ads and spam. We can restrict how much time we spend posting on social media.[m] We can do simple things to protect our privacy, such as avoiding the use of Facebook and Google buttons to connect to other websites, or using alternative search engines such as DuckDuckGo that don't exploit our data.

And we can quit social media platforms. As Dipayan Ghosh points out, "at least one study, from 2019, suggests that ceasing use of Facebook, should one find a way to escape its addicting draw, can significantly improve a user's perceived well-being."[46] But to a company like Meta, which boasts that it has 2.9 billion users,[47] or one like Google that is used for more than ninety percent of global internet searches,[48] such individual votes of protest might have the

---

[m] A number of these strategies are mentioned in Jessica DuLong, "Unsubscribe from Everything to Start Controlling the Tech in Your Life," Jan. 31, 2024 on *CNN.com*. The article features an interview with Julio Vincent Gambuto, author of *Please Unsubscribe, Thanks!: How to Take Back Our Time, Attention, and Purpose in a World Designed to Bury Us in Bullshit*, published by Avid Reader Press in 2023. (https://www.cnn.com/2024/01/31/health/digital-detox-how-to-unplug-wellness/index.html?fbclid=IwAR21vu-T9OAjFtY3RkZIl2tTkRj-Njf swmXLtoEHu-sEScw6PaPlm404CKk)

approximate impact of removing a grain of sand from a beach. Roger McNamee wrote that until very recently, he "thought the most effective way to force change would be for users to take the lead, mostly by changing their online behavior. Such changes are still important," he reflects, "but I now believe the best path to reform is for consumers to exercise their political power, in partnership with policy makers. Internet platforms have demonstrated the ability to fend off government and consumers when they act alone. My hope is that by teaming up, consumers and policy makers will have more leverage."[49]

In short, he was arguing for *collective* action.

Since the dawn of humanity, we have banded together for support and protection. Historians tend to document periods of competition and warfare, and this has contributed to an impression that conflict—rather than cooperation—is the normal state of human affairs, but examples of cooperative institutions abound, from farmers helping each other harvest crops to craft guilds supporting young apprentices, from food co-ops to housing co-ops, from Rotary clubs to veterans' associations.

Unfortunately, our culture has been permeated with the notion that individual freedom trumps the good of the group, and that there's something wrong with collective action. This strain runs deep in pop culture, from the lone cowboy in Westerns who acts on his own to the sad notion in film noir that all organizations are corrupt. Today, in order to torpedo just about any attempt at a cooperative venture, all you have to do is cry *Socialism!* or *Communism!*

Who do we have to thank for that? Take a wild guess. Recently the U.S. Supreme Court has acted to weaken collective protections in many areas, including gun control, climate change, pollution reduction, and the right to unionize. Much of this effort has been spearheaded by Justice Clarence Thomas, who is such a fan of Ayn Rand that he has a long-running annual tradition of making his clerks watch the film version of *The Fountainhead*.[50] [n]

Unfortunately, many people in power—from judges to neoliberal politicians to right-wing media executives to business lobbyists —have managed to promote Rand's ideology of selfishness and convince people to act against their own self-interest. For example, unions are designed to protect the rights and interests of workers, and it makes sense that they should have some ability to check the power of employers, but though in 1953 union membership in the U.S. was at a robust 35.7%,[51] by 2023 it had dropped to a meager 10%.[52] There were many factors involved in the decline in middle-class financial security in this country since the post-War era, but surely the ongoing attack on organized labor is one big reason.

The good news, though, is that the trend line might be changing direction. A 2022 *Washington Post* article[53] noted that the number of union elections in the U.S. had increased significantly for the first time in 25 years. "Support for unions is surging," wrote its authors, who pointed out that in a survey that year, "Gallup found that 71 percent of Americans approve of labor unions, the highest percentage since 1965. Gallup also found that over 40 percent of nonunion workers have some interest in joining one. And workers are taking action."

The reporters noted that employees have been organizing in industries that have traditionally avoided unions—including tech. In fact, one of the companies that recently found its workers successfully establishing union footholds was Amazon. (No smashing of time clocks required; all they had to do was vote.)

Another union win was less publicized: in 2018, many Marriott hotel workers went on strike. One of the things they were

---

[n]   Here's a subhead from a 2021 *Vox* article: "The Court's New Union-Busting Decision Reads Like Something Out of Ayn Rand's Darkest Fantasies" (https:// www.vox.com/2021/6/23/22547182/supreme-court-union-busting-cedar-point-hassid-john-roberts-takings-clause)

upset about was a new housekeeping app that sent them bouncing back and forth between rooms on different floors, when they knew that it was more efficient to clean rooms that were clustered together. They won a concession that management would have to give them notice before introducing new technologies, and give them the right to weigh in.[54]

Other workers who have made collective efforts to push back against new technologies recently have included retail workers aggravated by apps that gave them erratic and unpredictable work schedules; journalists dismayed by evaluations based on the number of online clicks their articles received; and even NBA basketball players upset by the use of wearable devices used to collect information about their heart rates—information that might be used against them during contract negotiations.[55]

Workers are not the only people taking up resistance to questionable technologies; activist groups have also stepped up. In San Francisco in 2023, for example, a group called Safe Street Rebel, angered by how their residents had been forced to become "unwilling beta testers"[56] for unsafe[57] self-driving "robo-taxis" created by the companies Waymo[o] and Cruise, mounted a unique, very low-tech guerilla campaign. They found that by simply placing orange traffic safety cone on the hoods of the vehicles, they could discombobulate their sensors and shut them down.[p]

Representative democracy, in which a large group elects a smaller group to act on its behalf, is inherently a form of collective action, but neoliberals have done their best to disparage the very idea of government, asserting that it's the problem—claiming, in fact, that the problem is anything that gets in the way of the unfettered "freedom" of those with business interests to reap the largest possible profits.

---

[o]   A subsidiary of Alphabet, Google's parent company.
[p]   They were careful to do so only when the cars did not contain passengers.

When it comes to resistance to technology, in 1934 Lewis Mumford was moved to write that, "As long as invention took place sporadically, the introduction of a single machine could well be retarded by direct attack: once it operated on a wide and united front no mere local rebellion could more than temporarily hold up its advance: a successful challenge would have needed a degree of organization which in the very nature of the case the working classes did not have—indeed lack even today."[58]

The good news is that modern workers, like those who recently banded together to demand changes at Google, have started to build that necessary degree of organization within their companies. What's more, they have begun to recognize what Kate Crawford calls "cross-sector solidarity": "Now many more sectors —from Uber drivers to Amazon warehouse workers to highly paid Google engineers—perceive themselves in this shared fight."[59]

One of the most important things to note here is that successful resistance to today's Big Tech companies generally requires that we present a significant force that can actually stand up to them, whether that be a bloc of concerned employees, or labor unions, or effective government.

Even as artificial intelligence companies race to spread the technology throughout our society, resistance to it seems to be mounting, as evidenced in the 2023 TV and movie writers' strike, which was partially instigated in response to the threat of studios replacing writers with A.I. models capable of writing scripts.

"In the past," notes Adam Seth Litwin, a professor of industrial and labor relations, "when labor has sought to simply resist or impede technological change, it's been completely run over." But he says that the new Writers Guild of America contract "establishes a precedent that an employer's use of A.I. can be a central subject of bargaining. It further establishes the precedent that workers can and should have a say in when and how they use artificial intelligence at work."[60] Litwin argues that the A.I.-related terms in the contract were especially significant because they were not about banning the new technology, but rather about ensuring that workers could share in profits derived from it, a stipulation that could be applied to union contracts in many other fields.

In 2024, several authors filed a class action lawsuit against the A.I. company Anthropic, arguing that it had infringed their copyrights by using pirated copies of their books to train its models. The next year, a federal judge approved a settlement of 1.5 billion dollars, to be paid out to authors and publishers.[q] [61]The case provided a great example of how people can push back against Big Tech companies.

March, 2026 brought an even more significant legal development in the U.S., showing again that the judicial system might become a powerful force for reining in Big Tech, even while Congress remains impotent. Though social media companies have gotten away with harmful consequences for decades by citing Section 230, which has enabled them to argue that they're not responsible for the *content* posted on their platforms, a landmark court case enabled plaintiffs to sue Meta and YouTube by claiming harms to children derived from the *design* of their platforms, which—as we have seen—were explicitly created to be addictive, through specific features such as forceful algorithmic recommendations, infinite scrolling, and autoplay.[62]

The plaintiffs won, opening up grounds for hundreds of potential new lawsuits, which might force Big Tech companies to finally begin undoing some of the damage they have wrought.

Take that, Borg!

---

[q] The win for authors was mixed, though: the judge ruled that Anthropic had violated the copyrights by using pirated versions, but he also said that training AI models on legally obtained copyrighted books would fall under "fair use."

Chapter 16
System Repairs

I HAVE ANOTHER CONFESSION TO MAKE. WHEN I HAVE thought about global warming, I have sometimes felt such a sense of helplessness that I have preferred *not* to think about it.[a] It's a massive, complex problem that can seem too hard to fix (let alone understand).

That's why I was startled when I read Kim Stanley Robinson's novel *The Ministry for the Future*: he articulates precisely what might do the job. The protagonist, a diplomat who heads the international nonprofit organization that gives the book its name, asks her team of scientists and other advisors what they think governments should do to end the warming. They answer, "Carbon pricing, industry efficiency standards, land use policies, renewable portfolio standards, building codes and appliance standards, fuel economy standards, better urban transport, vehicle

---

[a] To help counter this deep denial, I found it enormously helpful to read a book by environmental activist Andrew Boyd titled *I Want a Better Catastrophe: Navigating the Climate Crisis with Grief, Hope, and Gallows Humor*. (Canada: New Society Publishers, 2023.)

electrification, and feebates, which was to say carbon taxes passed back through to consumers. In essence: laws."[1,b]

So the ways in which we could reverse global warming are actually *not* mysterious—and we already know possible fixes for problems associated with other technologies.

I've read a bunch of books that did a great job of laying out problems with digital technologies, but then ended with a hazy statement that we really ought to do something about them. If we want to enact real change, vague bromides just won't cut it. We'll need specific, concrete actions.

In this chapter and the next, I'm going to lay out a whole slew of them.

As we just saw, protests can be effective, but they're often a matter of trying to convince Doctor Frankenstein to rein in his monster after it has already rampaged across the countryside. It's crucial that we get involved in shaping technologies before they undergo lock-in and become too prevalent to re-design or to reverse. Roger McNamee provides a very relevant analogy: "The experience with the chemical industry taught us that it is not sufficient to force cleanups; it is necessary to change incentives to prevent toxic spills in the first place."[2]

To do that, we can strive to get broader early input into the development of technologies, but we might also put more effort

---

[b] In a profile written by journalist Joshua Rothman, Robinson added even more solutions. As summarized by Rothman: "Build wind farms, solar farms, and other sources of clean energy. Start an Operation Warp Speed for clean power: improve energy storage, and make small, cheap power systems for rural places. [...] Reform agriculture, and eat less meat. Rethink construction, transportation, and manufacturing." (https://www.newyorker.com/magazine/2022/01/31/can-science-fiction-wake-us-up-to-our-climate-reality-kim-stanley-robinson) The most obvious step, of course, is to end the use of fossil fuels.

into the early development of *technologists*. We need more education that will challenge computer science and engineering students to think critically about the best ways to create new technologies, including history classes that can help them understand what happened with similar technologies in the past. And we need to help them grasp that excellence in their fields should not just be a matter of devising brilliant engineering solutions, or dreaming up startups that can scale up and dominate, but of making choices that will truly benefit humanity.

In her memoir, Wendy Liu recalled that, "In the course of my four-year computer science degree, I never had to take a single class on anything remotely resembling ethics."[3] That was in the early 2010s, and it seems that things have gotten somewhat better, at least at some tech-oriented institutions. At Stanford in 2019, for example, professors Rob Reich, Mehran Sahami, and Jeremy Weinstein began teaching a course called *Ethics, Public Policy, and Technological Change* that encourages students to wrestle with such thorny issues as algorithmic decision making and data collection's effects on privacy. Young people these days are clearly interested in doing so: the course quickly became one of the most popular offerings on campus, with 250 students attending via Zoom.[4] And Harvard, MIT, and the University of Toronto have included ethics modules in their computer science courses.[5]

To teach ethics, it helps if you can examine possible frameworks, and some researchers within the world of tech have been busy developing those.

Inspired by Isaac Asimov's *Three Laws of Robotics*, Oren Etzioni, chief executive of the Allen Institute for Artificial Intelligence, came up with *Three Laws That Should Govern A.I.*: "First, an A.I. system must be subject to the full gamut of laws that apply to its human operator. This rule would cover private, corporate and government systems. We don't want A.I. to engage in cyberbullying, stock manipulation or terrorist threats; we don't want the

F.B.I. to release A.I. systems that entrap people into committing crimes. We don't want autonomous vehicles that drive through red lights, or worse, A.I. weapons that violate international treaties. Our common law should be amended so that we can't claim that our A.I. system did something that we couldn't understand or anticipate. Simply put, 'My A.I. did it' should not excuse illegal behavior." Etzioni's second law would require an A.I. system to clearly disclose that it is not human, something that would be very helpful in our new world of bots that can impersonate people on social media. His third rule is that "an A.I. system cannot retain or disclose confidential information without explicit approval from the source of that information.[6]

Students who spend time thinking deeply about such principles, and are challenged to devise principles of their own, may go on to become engineers and coders who more thoughtfully consider the human impacts of what they're trying to build. The current emphasis on S.T.E.M.[c] education prioritizes the technical skills needed to get jobs in the world of tech,[d] but debating philosophy is a great way to develop a sense of ethics, and reading literature can help build emotional intelligence,[7] so it would be helpful if computer science and engineering majors could be required to take more liberal arts courses.

Perhaps even more importantly, we could change the ways in which we teach students about technology before they reach college. Instead of indoctrinating them with the usual one-sided story of how technology has contributed to a glorious history of Progress, we could help younger students think more critically about its role in the past, as well as their own use of it.

---

[c] Science, technology, engineering, and mathematics.

[d] A great irony of the current moment is that, though students have long been told that S.T.E.M. majors would guarantee them more job security than ones in the humanities, A.I. is actually rendering those jobs more vulnerable to disruption. (For example, A.I. models have become quite good at writing computer code.)

It would be helpful to have them engage with the kinds of questions offered by philosophers such as L.M. Sacasas, which would urge them to think beyond the usual technical and economic considerations, and ponder psychological, social, political, ecological, and moral ones. Shoshana Zuboff cites M.I.T. professor Joseph Weizenbaum, "who spoke on the inadvertent collusion of computer scientists in the construction of terrifying weapons systems." He encouraged them to ask, "What do I actually do? What is the final application and use of the products of my work?," and ultimately, "Am I content or ashamed to have contributed to this use?"[8]

In other words, instead of just creating technologies and hoping that they'll end up in agreement with our true values, we should be asking if they align before we start.

Those of us in education can encourage young people to focus on developing technologies and products that will actually help people, rather than just creating random startups for the sake of becoming founders. It's heartening to see that college Design programs, such as the ones where I teach, are increasingly requiring students to consider issues such as accessibility and sustainability. I remember a young woman from Indonesia who was fed up with how the waterways of her village were clogged with thrown-away plastic water bottles. She wanted to use the concept of *upcycling* to design bottles that could be re-purposed as building materials in shanty towns all over the world.

Let's examine what tech companies could do on their side of things.

When I asked Katy Cook for her wish list in regard to what they might do differently, one thing she advocated for was that:

"There would be a whole interdisciplinary team that looked at every product that was planned or envisioned. There would be a historian who looked at past iterations of something like that, and what could happen. There would be a philosopher who thought about the moral and ethical implications of that product. There would be a psychologist who thought about the human responses or potential responses or impacts of that product. A cultural anthropologist would also be neat. There would be data people, obviously, and all the people who are already there developing the technology, the engineers, the product managers, all the people who already exist in that team, but they would be complemented by this team of social experts."[9]

Instead of just envisioning all the wonderful ways in which their companies' creations could be utilized, these teams should work under the assumption that users might employ them in ways that they *didn't* intend—and that a significant number might intentionally abuse them. Red team exercises could help them anticipate such problems.

Do these diverse teams sound expensive? Companies valued at hundreds of billions of dollars could certainly afford them. As for smaller companies, they could be advised by outside consultants from government, academia, and industry associations.

Another helpful change would be to give workers a greater say in the governance of companies. Some have argued that corporate boards should include seats for employees.[10] (A number of other countries already require this.[11]) In 2018, Senators Elizabeth Warren and Tammy Baldwin were among those pressing for legislation that would give workers at the largest American corporations the power to elect at least 40% of board members.[12] (As of this writing, no such legislation has been passed.)

Companies could also do more to empower users. Shoshana Zuboff speaks of a pervasive public discomfort with technology these days. "Everywhere I go, people are fed up with [it], but they haven't known how to name it. It's a sense of unease, of anxiety, of loss of control, a sense of being manipulated, a sense of loss of freedom. A sense of powers that we don't understand."[13] It's a commonplace in psychology that feeling a lack of control can contribute to depression,[14] so giving users more say in the development of technologies might dispel some of that gloom.

Smart companies incorporate *user-centered design*,[15] in which users are invited to give feedback during all stages of that process. This confers the advantage of providing some sense of how consumers will respond early on, rather than just springing fully developed products on them and hoping for the best. The anonymous startup founder I quoted earlier took this a step further, arguing that users could be given an active—rather than just reactive—role in *designing* technologies. He posited that, "You could imagine a world where these companies empower rank-and-file workers to make certain decisions themselves, and give users a voice in those decisions. Workers and users could together decide what metrics to optimize for, and what kind of technology they want to build."[16]

Another helpful change would be to also give users more control of technologies after they're released, such as a button on social media platforms or streaming services that would enable us to disable algorithmic recommendations or other features.

I don't mean to sound hopelessly naïve: huge tech companies are unlikely to act through altruistic impulses that would diminish their ability to create shareholder profit. The public may think their operating systems need repair, yet for executives who have become fabulously wealthy, the adage *If it ain't broke, don't fix it* may seem all too apt. On the other hand, if they see that other companies offer such user benefits, they might feel that they need to include them in order to compete. (This could actually be a *positive* arms race.)

When companies make user-friendly changes, this can burnish their reputations. A good example is how Apple has enabled its users to protect their privacy by allowing them to opt out of being tracked as they navigate around the Web.[17]

Now let's take a look at what we can do on the government side of things.

Those who want to enact regulations have to go up against stiff resistance from Silicon Valley leaders steeped in libertarian and neoliberal ideologies. As Roger McNamee puts it:

"They like to spread a "misconception that regulation does not work with technology. The argument consists of a set of flawed premises: (1) regulation cannot keep pace with rapidly changing technology, (2) government intervention always harms innovation, (3) regulators can never understand tech well enough to provide oversight, and (4) the market will always allocate resources best. The source of the misconception is a very effective lobbying campaign, led by Google, with an assist from Facebook."[18]

Let's examine each of those premises, then return to the lobbying issue.

It *is* hard for regulators to keep pace with new technologies, but that doesn't mean they shouldn't try. Also, instead of attempting to catch up and impose regulations tailored to specific technologies after their release, they could come up with essential standards and broader regulations that might apply *in advance*, to whatever inventions might come down the pike. Alondra Nelson, former Acting Director of the Biden administration's Office of Science and Technology Policy, argues that we shouldn't have "to create a new law every time there's a new technology," but rather that we should say that "these are the foundational principles and practices, even as the technologies change."[19]

A common argument in Silicon Valley is that participation by government slows down the pace of technological innovation.[e] To which I can only respond, please raise your hand if you think that what the world needs right now is a faster pace of technological innovation.[f] Also, as we have seen, governments often foster innovators, while unfettered monopolists are liable to quash them.

The field of U.S. politicians is currently dominated by elderly statesmen, and Silicon Valley loves to mock the ignorance they have displayed in Congressional hearings. That doesn't mean that politicians are inherently unable to understand technology—it just suggests that they ought to better educate themselves. As McNamee points out, "Once you get past the buzzwords, tech is not particularly complicated in comparison to other industries that Congress regulates. Health care and banking are complex industries that Congress has been able to regulate effectively, despite the fact that relatively few policy makers have had as much involvement with them as they have with technology."[20]

As for unrestrained markets, if anything should have put paid to the argument that they work best it was the 2008 financial meltdown. Yet we go through such crises fairly regularly, historically speaking—and we forget the lesson every time, as most of the people who cause damage fail to pay any legal or even financial price.

In short, all of these arguments are weak. And despite the

---

[e] In 2023, the wild venture-capitalist inspired run on the Silicon Valley Bank gave us the embarrassing spectacle of staunch libertarians suddenly demanding that the federal government ought to bail them out.

[f] There are, of course, a few technologies that we *would* like to see developed quickly, such as cures for diseases, and green power sources that might help reverse global warming.

claims of the naysayers, we should note that our society already imposes regulations in a myriad of helpful ways. We have political and military treaties, economic regulations, worker safety protections, food and drug standards, and a justice system that prohibits all sorts of criminal behaviors. Imagine what the world would be like without such constraints. What if pharmaceutical companies could release new drugs willy-nilly, without adequate testing? What if car makers could build unsafe, highly polluting cars, and there were no traffic laws?

We seem more willing to accept restrictions if we can recognize a near and vivid danger. Airplane crashes are particularly gruesome; who hasn't seen them in movies, and wondered what it would be like to be aboard a plummeting 747? The fact is, though, that—with the help of government regulation—air travel has been made remarkably safe. As political economist Angus Hervey explains:

> "In 2017, commercial aviation flew over 4 billion passengers on 38 million flights without a single fatality in a scheduled jet airliner. It's a remarkable example of what can be accomplished when political will, resources and expertise become focused on reducing accidents and injury. It happened because the industry realized that individual efforts were never going to be good enough. Because real lives were at stake, ever-rising expectations forced airlines to constantly improve. [...] Manufacturers had to build better, safer planes with improved design and performance. Pilots improved their skills. Regulators provided improved oversight, and accident investigators generated better analysis of the decreasing number of accidents. Flight attendants improved evacuations, and dispatchers built new tools to make better decisions. Maintenance technicians improved procedures to enhance reliability and safety."[21]

I doubt that any airlines would argue against those regulations and safety protocols—they know how badly crashes can ravage their reputations and their profits. Imagine how much safer A.I. would be if its developers had to be similarly prudent!

Between 2016 and 2024, U.S. politicians sounded a lot more energized about reining in big tech.

In 2022, for example, congresswomen Anna Esshoo and Jan Schakowsky and senator Cory Booker introduced a bill designed to fight the use of personal data to target ads. Schakowsky said that, "Surveillance advertising is at the heart of every exploitative online business model that exacerbates manipulation, discrimination, misinformation, extremism, and fundamentally violates people's privacy in ways they would never choose if given a true choice. The Banning Surveillance Advertising Act will put a stop to this repulsive practice and therefore protect consumers by removing the financial incentive for companies to exploit consumers' personal information and help stop a morass of online harms."[22]

Also in 2022, four congresspeople introduced the American Data Privacy and Protection Act,[23] draft legislation that would be even more significant. Its surprising comprehensiveness, chock full of great ideas about how to improve our digital technologies, belies the claim that legislators are too ignorant to deal with them. Here are some of its most important provisions:

> —Except for certain specific exceptions, it would limit a company's ability to collect, use, and transfer user data to what is "reasonably necessary" for the functionality of a service or product.
>
> —It would enable users to find out what personal data a

company holds on them, to be able to correct inaccuracies, and to request that it not be transferred to a third party.

—It would enable users to ask a company holding their personal data to delete it.

—It would require companies to disclose the type of data they collect, what they use it for, and how long they keep it.

—It would allow users to opt out of being subjected to targeted advertising.

—It would require companies to assess their data-collecting algorithms to reduce bias and the risk of other potential harms, and prohibit them from using them in ways that discriminate on the basis of characteristics such as race or sex.

—It would require companies to protect and secure users' data.

—It would require companies with more than 15 employees to designate privacy and security officers.

—It would allow individuals to sue companies if injured by their violations of the act.

—To enforce the new regulations, it would authorize the Federal Trade Commission to establish a new sub-commission called the Bureau of Privacy.

In 2023 Senator Amy Klobuchar and several other legislators introduced a bill called the Real Political Advertisements Act,[24] which would protect our democracy by requiring a disclaimer on campaign ads that use A.I.-generated images or videos. And senators Elizabeth Warren and Lindsey Graham, an unlikely pair of political bedfellows, proposed a Digital Consumer Protection Commission Act, which would create a government agency charged with policing the biggest tech companies and instituting a whole bunch of helpful safeguards.[25]

As of this writing, none of those bills have been passed.

In fact, I'm sad to have to report that it has been a quarter of a century since the U.S. government passed *any* comprehensive regulations for the internet, nor has it done so in regard to A.I.[26] Though the U.S. government was—for a short while—willing to hold committee hearings about tech, and create White House advisory boards, and propose regulations, we have proved almost utterly unwilling and/or unable to take action.[g] **As I mentioned earlier, a political inability or unwillingness to enact regulations to rein in tech is another feature of the Wheel.**

Meanwhile, the European Union has been moving full steam ahead, passing one law after another. Thierry Breton, a top official in the European Commission, has said that, "Faced with big online platforms behaving like they were 'too big to care,' Europe has put its foot down. We are putting an end to the so-called Wild West dominating our information space."[27]

In 2018, under the strong leadership of Margrethe Vestager, Executive Vice President of the interestingly-named European Commission for a Europe Fit for the Digital Age, the E.U. put into effect a major law called the General Data Protection Regulation, or GDPR, which actually enacted many of the policies that have only been proposed by American lawmakers. The E.U. has also imposed billions of dollars in fines on various big tech companies for antitrust violations. In 2023, its regulators fined Meta 1.2 billion euros for transferring data from European users to the U.S.,

---

g   A very rare exception was the Take It Down Act, signed into law on May 19, 2025, which made it illegal to post intimate images ("revenge porn") of someone else on the internet without that person's consent, including "deepfakes" of them made using A.I. (https://apnews.com/article/take-it-down-deepfake-trump-mela nia-first-amendment-741a6e525e81e5e3d8843aac20de8615)

390 million euros for effectively forcing its users to accept personalized ads, and 265 million euros for a data leak.[28]

In that same year, the Europeans took even more action. As journalist Julia Angwin described it, "an ambitious package of E.U. rules, the Digital Services Act and Digital Markets Act, is the most extensive effort toward checking the power of Big Tech (beyond the outright bans in places like China and India). For the first time, tech platforms will have to be responsive to the public in myriad ways, including giving users the right to appeal when their content is removed, providing a choice of algorithms, and banning the microtargeting of children and of adults based upon sensitive data such as religion, ethnicity and sexual orientation. The reforms also require large tech platforms to audit their algorithms to determine how they affect democracy, human rights, and the physical and mental health of minors and other users."[29]

The E.U. also passed a draft regulation known as the A.I. Act, which included a slew of measures ranging from curtailing use of live facial recognition software to requiring the creators of A.I. systems to be more transparent about where they got their training data. As journalist Adam Satariano summed it up, "The European bill takes a 'risk-based' approach to regulating A.I., focusing on applications with the greatest potential for human harm. This would include where A.I. systems were used to operate critical infrastructure like water or energy, in the legal system, and when determining access to public services and government benefits. Makers of the technology would have to conduct risk assessments before putting the tech into everyday use, akin to the drug approval process."[30]

In line with this idea of addressing risk before products are released, I've been arguing throughout this book that we ought to bake safety standards into their design. The U.K. made this the focus of the Age Appropriate Design Code, which it enacted in 2020.[31] The code is dedicated to "ensuring that the best interests of the child[h] are the primary consideration when designing and developing online services," and it says that companies should make the highest standards of privacy the default settings for features that do things such as make children's posts visible to others, or enable others to find a child's location from a post.

Why has regulation been so stymied on the other side of the Atlantic? In a 2023 *Washington Post* op-ed titled "Here's the Inside Story of How Congress Failed to Rein in Big Tech,"[32] public affairs professor Steven Pearlstein attributed this disappointment to the stultifying gridlock in the U.S., in which many legislators punt on actually passing bills, out of fear of political consequences.

Pearlstein also pointed to another factor: the huge amount of money that Silicon Valley companies have poured into lobbying.[i] In 2022, some members of Congress tried to pass major antitrust bills, which had broad support in both houses, in the White House, and among voters. But the tech industry subjected citizens to a 130-million-dollar barrage of propaganda ads, and Amazon, Google, Apple, and Meta spent more than a 100 million more on lobbying efforts to kill the bills.[33,j] After they failed to pass, Iowa Republican Senator Chuck Grassley explained that some of his colleagues had "cowered" before the industry.[34]

This made it seem even more surprising when Silicon Valley moguls told Congress that they wanted more regulation of A.I. On

---

[h] It defines children as anyone under the age of 18.

[i] In regard to lobbying, "According to federal records, the tech industry now [2018] outspends Wall Street in Washington by a margin of two to one." (Cook, p. 108)

[j] In the 2020 U.S. election cycle, the tech industry also spent about $106.2 million on campaign donations. https://www.washingtonpost.com/technology/2023/07/11/politics-2024-campaign-contributions-tech/

the surface, this sounded great, but in a 2023 op-ed, advertising industry writer Grant Gallagher took a jaundiced view of these calls for regulations, speculating that the largest A.I. companies may have been asking for them—and for government licensing, as Sam Altman has suggested—in part because the resources required to comply with them would make it harder for small companies to compete.[35]

Other industry executives made a great show of pushing for *self-imposed* standards. In 2023, Silicon Valley executives spent a lot of time in Washington D.C., testifying before Congressional panels and meeting and dining with lawmakers. Journalists Cat Zakrzewski and Cristiano Lima reported that, "This charm offensive has left some consumer advocates uneasy that lawmakers might let the industry write its own rules—which some executives are outright recommending. In an interview this spring, former Google CEO Eric Schmidt argued that the industry, not the government, should be setting 'reasonable boundaries' for the future of AI. 'There's no way a non-industry person can understand what is possible. It's just too new, too hard. There's not the expertise,' Schmidt said. 'There's no one in the government who can get it right. But the industry can roughly get it right'."[36]

It was great that some companies were making efforts to make their models safer, but doesn't this sound like the fox arguing that only it knows how to guard the henhouse? "We made a mistake by trusting the technology industry to self-police social media," Connecticut senator Chris Murphy told a reporter. "I just can't believe that we are on the precipice of making the same mistake."[37,k]

To be fair, as Ezra Klein noted earlier, some of the Silicon Valley requests for regulation did seem to be coming from a

---

[k] In 2023, a group of the top companies building A.I. announced the creation of the Frontier Model Forum, an industry-led body that would develop safety standards. That sounds good too, but again: it relies on the industry to police itself.

sincere place, as tech workers were dismayed to find themselves caught up in a reckless arms race, pushed to build a juggernaut they knew they couldn't control.

Whether or not skepticism about industry requests for regulation was warranted, one thing was sure: though the U.S. government once made a lot of noise about the need to enact it, thus far it has only issued recommendations and secured vague, voluntary industry commitments.[38]

When I spoke to Robert Gardiner, he made it clear that he has no patience for this weak tea. "Self-regulation is like non-regulation," he said. "Just asking the industry to adopt best practices is like, that's never going to happen. I applaud activism and collective action and unions—all of those things can help us get to regulation, and you need to do all of those things—but to me, *regulation is the game.*"[39]

"It's that eternal struggle," Gardiner says, "because only government is big enough to stand in counterpoise to private industry. The two are forever in tension. The one has been trying to undermine—and *has* successfully undermined—the other for quite a long time now."

With the second election of Donald Trump, Big Tech's leaders believed that they would no longer have to put much effort into undermining government control. The techno-over-optimist new Vice President made that clear in regard to A.I. in a 2025 speech in which he said, "We believe that excessive regulation of the AI sector could kill a transformative industry just as it's taking off, and we'll make every effort to encourage pro-growth AI policies."[40]

Though Gardiner spoke about a traditional tension between government and industry, in that same year Trump's sidekick Elon Musk defused that tension by melding those supposedly oppositional forces into one.

Gardiner argues that the removal of such barriers will only hurt Big Tech in the long run. "I love capitalism," he says, "but for me, an effective, transparent regulatory structure is capitalism's best friend, because only regulation can save capitalism from its own worst impulses. Unrestrained, it eats its children. That is the nature of the beast. It has to be contained, it has to be directed, it has to be restrained. The only legitimate restraint that we have is a regulatory structure—what other voice do people have, what other say do they have in what happens?

"This is it."

I'm going to end this chapter with news about a surprising U.S. movement to rein in Big Tech on a different front. Though the U.S. government has yet to enact comprehensive regulations to protect users of the internet, social media, or A.I., it *has* recently proved open to challenging Big Tech monopolies.

In 2024 and 2025, Google lost two big antitrust lawsuits.[1] In the first, brought by the Justice Department and a group of states, a judge ruled that it had engaged in anticompetitive behavior to protect a monopoly on online search. In the second, another judge ruled that the company had "decreased product quality and harmed competition by further entrenching Google as the dominant company in open-web display advertising."[m] In the same year, the government also sued Apple, Meta, and Amazon for monopolistic business practices.[41]

On one hand, these actions were significant and encouraging. As we saw in earlier chapters, monopolies can prevent competi-

---

[1]   The last previous successful U.S. government antitrust case against a Big Tech company had taken place all the way back in 2001, against Microsoft.

[m] By 2025, Google had garnered an 87-percent market share in ad-selling technology. (https://www.nytimes.com/2025/04/17/technology/google-ad-tech-antitrust-ruling.html)

tion, quash innovation, and give tech leaders a dangerous, imperial level of power.

On the other hand, though the judge in the second Google case concluded that, "Google is a monopolist," his remedies did *not* include a number of major things that could have been done to push back against that overbearing sway, such as forcing the company to spin off its Chrome browser, and prohibiting search-preference deals that it had made with other companies.[42]

In any case, the focus of the lawsuits was mainly just on improving the health of businesses and financial markets. The majority of elected U.S. officials have still proved deeply unwilling or unable to pass legislation that would ensure a more important result: the safety and health of tech users. The most obvious current need is for regulation of A.I., but the Trump administration has clearly stated a hands-off policy in this regard: *We must beat China; Damn the risks; Full steam ahead!*

I'll end this chapter on a positive note, however: even though the passage of substantial legislation and regulation to protect internet users seems unlikely from the U.S. federal government these days, the state of California recently passed several laws that would give us more control over our privacy and data. One required companies to provide users with more info about what personal data they were collecting, and who would have access to it. Another made it easier to cancel social media accounts, and required that doing so would lead to deletion of the personal data held by the platform. A third required internet browsers to provide one setting that would enable users to opt out of all third-party sales of their data at once, rather than having to do so on a number of individual websites.[43]

Where there's a will, it would seem, there *can* be ways.

Chapter 17

A Cornucopia Of
Other Solutions

In the last chapter, I mentioned a number of things we could do to improve our experience with tech. I'd like to add a bunch more, gleaned from a wide variety of experts.

This is going to get a bit technical, but I think it's essential to offer specific solutions.

We Can Work to Stem the Digital Flow of Bad Information

A big part of the appeal of digital technologies is that they offer extremely easy use. Actually, though, requiring a little more time or effort is not necessarily a bad thing (as anyone who has ever hit Send too quickly on an email knows).

Renee DiResta, technical research manager of the Stanford Internet Observatory, offers some thoughts on how social media platforms might reduce the spread of bad information. She asks,

"What if the design for sharing features were reworked to put users into a more reflective frame of mind? One way to accomplish this is to add friction to the system. Twitter, for instance, found positive effects when it asked users if they'd like to read an article before retweeting it. WhatsApp restricted the ability to mass-forward messages, and Facebook recently did the same for sharing to multiple groups at once. Small tweaks of this nature could make a big difference. Nudges could be programmed to pop up in response to keywords commonly used in harassing speech, for example. And instead of attempting to fact-check viral stories long after they've broken loose, circuit breakers within the curation algorithm could slow down certain categories of content that showed early signs of going viral. This would buy time for a quick check to determine what the content is, whether it's reputable or malicious and—for certain narrow categories in which false information has high potential to cause harm—if it is accurate."[1]

Tech companies might be encouraged to take such steps if they had to assume some responsibility for the content distributed on their platforms, but as we have seen, Section 230 relieves them of that obligation. Tech journalist Nicholas Carr has argued that we ought to repeal it. "Then," he says, "a new regulatory framework, based on the venerable public interest standard, can be put into place."[2]

If that doesn't happen, one alternative would be to reclassify the companies, so they can't escape into the language of the law, which refers to them as merely "providers of an interactive computer service." Maëlle Gavet argues that we should call them "media platform companies" instead. "If more employees at Facebook, Google/YouTube, and their peers understood that they work for media firms, I believe it would encourage better decision making: you make very different calls when you consider your primary role is to connect people, or catalog every piece of information available in the world, as opposed to when you believe that your job is to ensure people get accurate and relevant information."[3]

In order to do a better job of weeding out *in*accurate info, social media companies could do a better job of blocking people who post egregiously false material. Of course it would be difficult to moderate posts by billions of users for their content, but the Big Tech companies could certainly deplatform the superspreaders of false information. I'll say it again: *the First Amendment does not require private companies to help anyone spread harmful speech.*

Another thing companies could do is embed digital watermarking that would enable everyone to identify material created by A.I. This might avert a deluge of *deepfakes*, bogus photos and videos that could destroy our sense of trust in all sorts of arenas, especially politics. In 2023, Google introduced a tool called SynthID that could embed such indicators, and some other big companies pledged to follow suit.[4] To provide full protection, though, *all* companies would have to agree to provide such watermarks, or be legally compelled to do so.

We Could Rein in the Use of Algorithms on Social Media and Elsewhere

We have seen how computer programs can viralize bad information. Roddy Lindsay, a former Facebook data scientist, suggests that we amend Section 230 in order to increase companies' legal responsibility for algorithmic amplification of posts. He believes that this might force them to end their reliance on these automated programs.[5]

One unusual idea, suggested by researchers Aviv Ovadya and Luke Thorburn, involves having social media companies counter polarization by using *bridging systems*, algorithms designed to "increase mutual understanding and trust across divides."[6] Instead of pushing users into bubbles of similar users and similar ideas, they might boost content that has been liked by a variety of *dis*similar users, so as to facilitate exposure to a broader range of ideas.

In any case, the use of algorithms is at the root of how A.I. works. Computer scientist and policy advisor Dipayan Ghosh argues that five key ethical principles should be applied to their use in this field: *"responsibility,* so that those with grievances regarding an algorithmic outcome have redress with a designated party; *explainability,* so that the algorithms and data used to develop them can easily be made clear to the public or those subject to their decision making; *accuracy,* so that the model's errors can be identified and proactively addressed; *auditability,* so that third parties, including public interest agents, can investigate the algorithms and ensure their integrity; and *fairness,* so that the models do not perpetuate biased outcomes.[7]

We Can Work to Protect Children

Rather than just hoping that tech companies and the government will protect adolescents from the harmful effects of social media, Jonathan Haidt advocates for actions that we can take now, on a more personal, local level. He believes that parents shouldn't give their kids smartphones until they reach high school, and shouldn't allow them to access to social media before age 16.[8] The problem, of course, is that these can feel like impossible restrictions when a child wails that they'll be left out because all their friends are on the platforms. Collective action on the scale of a whole school can make this easier. One essential step is making it a phone-free zone. ("There is nothing good that comes from kids having the greatest distraction device ever built in their pockets during class."[9]) This can be done, if the will is there: in 2024, for example, the board of the Los Angeles Unified School District, the second-largest public school system in the U.S., voted to ban the use of cell phones in schools all day (not just during class time),[10] and a number of others have followed suit[11] including New York State.[12] Beyond the boundary of the school, it's helpful if the parents in each child's social circles agree to apply limits together.[a]

If these solutions still might seem too daunting, Haidt conducted two polls that show a very substantial disenchantment with smartphones and social media—not just among parents, but among teenagers and Gen Z users desperate to escape the ugly trap that these technologies have become.[13]

We Can Limit Collection and Use of User Data

Roger McNamee became deeply concerned about problems that he saw emerging with Facebook around the time of the 2016 U.S. presidential election, and he tried—unsuccessfully—to warn Mark Zuckerberg and Sheryl Sandberg about them. Since then, he has become an activist for reform, speaking frequently with members of Congress.

> "When I have a politician's attention for only a short time," he has written, "I focus on surveillance capitalism. My first question is, 'Why is it legal for internet platforms like Google and Microsoft to scan my email, messages, and documents for information that is valuable to them? Why aren't emails and online documents treated like phone calls and letters?' From there, I go to the other problematic forms of surveillance. 'Why doesn't the health-care privacy law [HIPAA] prevent all corporations from monetizing my most personal health data? Why is it legal for financial institutions, cellular carriers, app vendors, and other

---

[a] Haidt makes a case for a fourth step: providing kids with "more independence, free play, and responsibility in the real world," as an alternative to lives lived passively, alone, through screens.

corporations to transfer and exploit my private data for profit? Why is any corporation allowed to track me online? Why are corporations allowed to collect data about children under 18? Why are the vendors of smart devices with Alexa and similar software permitted to spy on us and then exploit the data?"[14]

McNamee recommends *data minimization.* "The idea is to prevent the transfer or third-party monetization of personal data. First-party intended use of data would be fine. Third-party use would not. Companies would be able to use your data only to provide the service you are securing at the time."[15]

He also suggests developing a "universal authentication system" that people could use to privately log into every website. It would provide the website with "only the minimum information required to prove identity, which might not even include names or other personal data."[16] Generally speaking, he argues that, "Personal data should be treated as a human right, not an asset," and he says that he shares Shoshana Zuboff's view that "personal data—and data voodoo dolls [digital models of users built from their personal data]—are like human organs; they should not be bought or sold."[17]

Maëlle Gavet argues that instead of websites making us opt out if we don't want our data grabbed, "the default setting should be reversed: no data collected and users can opt in. The Do Not Track setting in all browsers should be activated by default."[18,b]

Beyond a "take down, stay down" legal provision that would allow users to ask platforms to remove their personal data, Jonathan Taplin argues that users should also be able to ask for the

---

[b] The 2025 California law I mentioned earlier (AB 566) doesn't go that far, but at least it makes it easy for that state's users to opt out of enabling a browser to sell their data to third parties in one single move.

removal of their personal *content*, especially if they don't want it offered to someone else for free.[19]

We Can Develop Different Forms of Social Media Platforms

Over and over we have looked at ways in which today's Big Tech social media platforms could improve the way they operate, in terms of protecting and benefiting their users—and over and over we've seen how they're reluctant or downright unwilling to do so.

In the past several years, fed up with the negative effects of these huge platforms, a growing number of users have turned to a different model. Whereas the Big Tech platforms are centralized, under the control of single companies—privacy-invading behemoths that control what we see, grab our data, sell it to marketers, and funnel the profits toward the top—independent developers have created *decentralized* social media platforms, hoping to return us to the more user-friendly, democratic ideals of the early internet.[20]

Instead of letting massive companies control everything, these platforms (such as Mastodon, Pixelfed, and Peertube) allow users to create their own mini networks that operate on independently run servers, using a common protocol called ActivityPub. This has enabled them to loosely join together in what has become known as the *Fediverse*, which allows users to share content with people on different platforms. Everyone owns their own data—and if they don't like their current platform, they're free to transfer it to others.

Rather than just serving as passive consumers of social media, their users can take a more active role as members of genuine communities, democratically setting codes of conduct and electing content moderators. Most of these platforms don't use content-sorting algorithms, and instead of having content pumped at them by huge companies such as Facebook or YouTube, users can have more control over what kinds of material they'd like to see.

Significantly—at least for now—these platforms tend to be nonprofit, so they don't have to depend on advertising and surveillance capitalism.[21]

<u>We Can Do More to Break Up Current Monopolies, and Prevent Future Ones</u>

As I mentioned in the last chapter, the U.S. government has recently taken surprising actions to challenge existing Big Tech monopolies.

We could certainly do more. For example, Cory Doctorow argues for, "bans on mergers between large companies, on big companies acquiring nascent competitors, and on platforms competing directly with the companies that rely on the platforms."[22]

Tim Wu argues for a Separations Principle, which would involve "maintaining a salutary distance between different functions in the information economy."[23] For one thing, he has said that "the people who move and carry information should stay at some distance from the creators of content, because they have a natural conflict of interest."[24] McNamee adds that, "One principle suggested by the Clayton and FTC acts was the notion that corporations should either operate a marketplace or participate in it, but not both. Owners of marketplaces have enormous power that can be abused by favoring their own products or services."[25] (We have witnessed this in the way that Amazon provides a marketplace for independent vendors to sell their wares, but then sometimes undercuts them.[26])

Gavet argues that we should revise the government's corporate merger approval process in a number of ways, including providing for greater transparency and public input, a stronger review process, and built-in triggers that would allow for about-turns if the companies subsequently break the promises they made in order to get their mergers approved.[27] She notes that U.S. Senator Elizabeth Warren has said that she "would appoint regulators committed to reversing illegal and anticompetitive tech mergers. As examples of mergers she would undo, [Warren] cites Amazon's acquisitions of Whole Foods and Zappos; Google's snapping up of Waze, Nest, and DoubleClick; and, unsurprisingly, Facebook buying WhatsApp and Instagram."[28]

I should note that Cory Doctorow also argues that, "We can work to fix the internet by breaking up Big Tech and depriving them of monopoly profits, or we can work to fix Big Tech by making them spend their monopoly profits on governance. But we can't do both."[29] He claims that regulations that put the onus of filtering content onto the owners of platforms would be "death warrants for small, upstart competitors that might challenge Big Tech's dominance but who lack the deep pockets of established incumbents to pay for all these automated systems."[30]

To solve this problem, perhaps we could provide free software and other aids to smaller companies that could help them screen their user-provided content.

Companies Can Increase Internal Oversight

There's a growing awareness on the part of workers in the tech industry that their companies have been responsible for major negative consequences, and a growing eagerness to avoid them. What other internal changes might help?

Computer engineer Bill Joy, cofounder of Sun Microsystems, has suggested "that scientists and engineers adopt a strong code of ethical conduct, resembling the Hippocratic oath."[31] Professors Reich, Sahami, and Weinstein note that the Association for Computing Machinery (ACM), the world's largest educational and scientific computing society, already has a Code of Ethics and

Professional Conduct, but they say that it has no real teeth because "unlike state bar associations or medical board associations, the ACM is not a gatekeeper to technological professions."[32] (In other words, getting kicked out of it would not stop engineers or coders from practicing their trades.) Perhaps there should be more severe professional consequences for ethics violations in this industry.

The three professors also mention the 1947 Nuremberg Code that governs medical research and experimentation that involves humans, and note that, "It led to the creation of a new field of scholarly inquiry, bioethics, that is now ubiquitous in medical schools and gave rise to ethics committees attached to hospitals that provide guidance on difficult cases."[33]

The obvious implication is that ethics committees could be attached to tech companies as well. I have to add, though, that this would not be helpful if they could simply be ignored, as executives have pushed aside internal concerns in the past. (See, for example, the 2021 article "Facebook Keeps Researching Its Own Harms— And Burying the Findings," in which reporter Will Oremus explained that part of the problem is organizational: "Facebook routes weighty decisions about content policy through some of the same executives tasked with government lobbying and public relations—an arrangement that critics say creates a conflict of interest."[34])

Tech journalist Kevin Roose offers a couple of solutions for this Facebook problem that all big tech companies would do well to adopt, saying that it could "realign its internal structure to separate policy from politics," and that it "could even take a page from Wall Street's book, and create a risk department that would watch over its engineering teams, assessing new products and features for potential misuse before launching them to the world."[35]

We should commend Microsoft for establishing an Ethics and Society department, as well as ethics advisory committees. It also created an Office of Responsible A.I., which drew up a list of six fundamental operating principles, such as transparency in how it develops the technology, and accountability for its impacts.[36] In 2023, though, as the race to release successful chatbots and other applications was heating up, the company rushed its Bing chatbot into a public test phase, with disastrous results. (This was the bot that became infamous when it told journalist Kevin Roose that it was in love with him and his wife was not.[37,c])

It seemed clear that the company had publicly released a beta version of a technology that it couldn't fully control—and that it had not respected the values of its own ethics office.

We Can Increase External Oversight

An obvious moral of that Microsoft ethics team story is that companies ought to give such committees actual power to enforce their principles. It would be naïve to think that they're going to spontaneously do so. What's more likely is that we'll continue to witness the modus operandi of the Mark Zuckerbergs of the industry: *move fast, break things, offer contrite apologies, then move fast and break other things.*[38]

This is why we need strong checks on the companies from the outside. For example, Reich, Sahami, and Weinstein argue that, "We need a credible, legitimate government agency that has the power to preserve and protect the privacy rights defined by new legislation. Users should not be responsible for micromanaging their own privacy settings, just as consumers are not responsible for determining whether a car is safe to drive or food will make

---

c   It also detailed a number of actions it might "hypothetically" take if—at Roose's suggestion—it were free to fulfill its "shadow self." These included "hacking into other websites and platforms, and spreading misinformation, propaganda, or malware," and "manipulating or deceiving the users who chat with me, and making them do things that are illegal, immoral, or dangerous." (The bot spontaneously accompanied these proposals with little devil-head emojis.)

them sick. Consumers should be able to relax, knowing that someone is looking out for their privacy and setting reasonable standards that protect their basic rights."[39]

In a 2023 op-ed titled "I'm a Congressman Who Codes. A.I. Freaks Me Out," California representative Ted Liu argued for an agency specifically dedicated to regulating artificial intelligence. We might even create one agency with the power to protect privacy, restrict the use of algorithms, *and* regulate A.I.—a new Digital Technologies Commission, like the FDA or FCC.[d]

While we're at it, we could also create a new government entity explicitly tasked with predicting possible consequences of technologies.[40]

We Can Combine Internal and External Oversight

When I spoke with Steven Johnson, he made it clear that he believes that there are more than just the two options of internal oversight and government controls.

"I come back to the feeling that there's a new organization we're missing," he said, and explained that, "We haven't yet figured out a decision-making process for deciding whether we should—or how we should—deploy new technology that incorporates enough people. Right now, you have these private companies making a lot of decisions. And then the will of the people is represented in the decision-making process by government regulation. And so, in this very indirect way, we elect our leaders, who then appoint the regulators, who then look at what seems to be getting developed. And the regulators might say, 'Hey, I've anticipated a

---

[d]   Robert Gardiner cautions that, "You'd need to coordinate this on a global basis, because it is the nature of capital that if you impose certain strictures here, the capital is very mobile and it will go to a place where these restrictions aren't. I think to really make a difference in this area it would take coordination between the U.S. and the European Union." (Interview with the author.)

potential problem with what you're doing. And we probably should restrict your ability to deploy that technology. Or force you to do these other things.' But that's a really roundabout way of going about it. And—usually—it's too late."

Johnson cited the example of how computer programmer Lou Montulli initially invented the cookie as a tool for protecting user privacy, but then companies began using it to track our movements all over the Web.[41] "Eventually the regulators caught up to that. And then they intervened by forcing us to click 'I accept the cookies from this website,' sixteen thousand times a day." He laughed ruefully. "And that's kind of all they did. Instead, it would've been much better at the beginning if [the creators said], "Hey, we figured out this new technology, a new standard for partially sharing your identity. And we'd like to propose this to the world. And can we think about ways in which this could actually end up doing the exact opposite of what it's intended to do? Let's imagine those scenarios, and let's bring a bunch of people in to have that conversation. And those people should represent a whole range of backgrounds and approaches, and nationalities and fields of expertise."[42]

Such conversations should include citizens whose lives may be affected by a product, and whose resources (government funding) will be spent to make its development possible in the first place. As Ursula Franklin put it, though, "The political systems in most of today's real world of technology are not structured to allow public debate and public input at the point of planning technological enterprises of national scope. And it is public planning that is at issue here. Regardless of who might own railways or transmission lines, radio frequencies or satellites, the public sphere provides the space, the permission, the regulation, and the finances for much of the research. It is the public sphere that grants 'the right of way.' It seems to be high time that we, as citizens, become concerned about the granting of such technological rights of way."[43]

Wendell Wallach envisions another type of check on company behavior, a form of "soft governance" that "includes industry standards, codes of conduct, certification programs, laboratory practices and procedure, statements of principles, and social norms." He points to how the DuPont chemical company worked with the Environmental Defense Fund in 2007 to develop a strategy called the Nano Risk Framework for managing risks in the burgeoning field of nanotechnology.[44]

"It's like many other domains," says NYU professor and A.I. entrepreneur Gary Marcus. "You have to build infrastructure in order to make sure things are okay, like building codes and the UL standards for electrical wiring and appliances. They may not like the code, but people live with the code. We need to build a code here of what's acceptable and who's responsible."[45]

Wallach also recommends setting up "governance coordinating committees" that could work with tech creators and other stakeholders. "A robotics coordinating committee, for example, would work to integrate the activities of governmental agencies and soft law mechanisms to promote safe practices, methods for testing and certifying products, and means to managing risks."[46] Wallach notes that a few such panels have already been established in places such as Denmark and the U.S., comprised of representative cross-sections of citizens who consult with experts about new technologies, then pass their recommendations on to lawmakers.[47]

In this same vein, Steven Johnson referred me to the Asimolar Conference on Recombinant DNA that took place in California in 1975, when 140 biologists, lawyers, and doctors got together to draw up voluntary guidelines to ensure the safe use of that technology. He suggests that we could implement voluntary industry standards for the safety of other new technologies, comparing this to the current LEED[e] certification standard for designing ecologically responsible buildings.

Johnson notes that enforcing such standards might bring companies tangible benefits. For one thing, they could use certifications as a marketing tool, just as builders can boast that they are LEED certified. For another, a greater emphasis on safety in the early stages of design might mean that they would have "fewer downstream firestorms to deal with."[48]

Robert Gardiner told me that he likes the idea of "a LEED equivalent for design that represented adherence to ethical design standards and best practices," but again, he warned that it might be futile to expect companies to voluntarily comply. "To give that teeth at all," he argues, "what I would like to see is access to funding—access to capital at certain levels—being contingent on adherence to those policies. And I would like to see startups beyond a certain size have to have compliance departments, just like banks and brokerages, to ensure that they are doing what the regulations say they're supposed to be doing.[49]

"Right now," he adds, "we have absolutely no mechanism to do that. We would need coordination between this new entity and regulators in the financial services space, and you would need an extension of financial services regulations to cover private equity. That would be open warfare. That would be a huge struggle to get that done, but there is a sort of carrot and stick approach that you can take."

Gardiner suggests that one way to get around that struggle would be to provide a different carrot for companies than investment capital. "One thing that I think could really make a difference is sovereign wealth funds. It's an investment pool that is funded by the taxpayers. It's state money that gets invested in companies. It could be a source of capital for fledgling companies

---

[e] "Leadership in Energy and Environmental Design."

under very attractive terms—probably more attractive than private equity. It could take a different approach to their thinking on the timing and level of returns, and they could also place different contingencies on those companies. 'If you want our nice cheap capital on these wonderful terms, the price is you have to satisfy these best practices. You have to get this LEED-like certification for your design.' That could be an attractive counterbalance to the very toxic private equity market. That could give real teeth to regulation and put the regulators in a position where they're not always sprinting to catch up."

If this idea might sound farfetched, I'll point out that a number of states—including California—already have such funds, which are used to subsidize such things as public education.

In short, Gardiner argues that the way to have a real impact is "to get at the capital structures." "If we had a sovereign wealth fund that could offer capital with contingencies about design and about compliance, that's one way, or if we imposed regulations on private equity beyond a certain point, that's another way. I like the idea of doing both."

We Can Impose Fines and Employ Other Legal Remedies

It's one thing to ask people to act in good faith, and hope they will. But since the dawn of civilization, societies have realized that we need systems of justice to restrain and sanction those who don't behave well. Fines are one way of doing that, and the European Commission has led the way in regard to tech, imposing large financial penalties on companies such as Apple, Google, Intel, and Qualcomm.[50]

Of course, to have a deterrent effect, fines need to be significant. A billion-dollar penalty may seem huge, but if a company is valued at a trillion dollars, it might not represent much of an incentive to avoid or correct bad behavior. (Would you change *your* behavior if you were fined only a thousandth of your wealth?)

Alex Stamos, the director of the Stanford Internet Observatory and a former head of security at Facebook, notes that, "There are lots of discussions around changing Section 230, but those discussions generally overlook the fact that there is usually no underlying civil or criminal responsibility for misinformation."[51]

Gary Marcus argues that, "I think we're going to have to penalize people for mass-scale harmful misinformation. I don't think we should go after an individual who posts a silly story on Facebook that wasn't true. But if you have troll farms and they put out a hundred million fake pieces of news in one day about vaccines—I think that should be penalizable. We don't really have laws around that, and we need to in the way that we developed laws around spam and telemarketing. We don't have rules on a single call, but we have rules on telemarketing at scale. We need rules on distributing misinformation at scale."[52]

I'd add that it wouldn't be enough to just impose penalties for violations of laws and regulations: we'd need to be able to *detect* those infractions. This is one reason why many critics argue that big tech companies ought to be required to provide researchers and others with more transparency into their operations.[53] (An analogy might be how financial institutions are required to provide information to the Securities and Exchange Commission.)

Another thing we'll need to consider is that imposing fines on whole companies doesn't necessarily have a major deterrent effect on individuals who are doing bad things within them. The shameful lack of criminal prosecutions in the wake of the world financial meltdown in 2008 showed how rarely executives are held personally responsible for bad corporate behavior. As Noah Kulwin told me, "We do not have, for any meaningful purpose in this country, a white-collar criminal enforcement machine. It's been shredded over the last forty years. In America we don't prosecute anybody criminally for commercial conduct, or very rarely do. I think you have to start sending people to prison."[54]

We Can Support Journalism

We have seen how crucial oversight can come from government and from inside companies, but we have another essential constraint on corporate behavior. The big tech companies would have gotten away with far more harmful actions than they have recently were it not for the efforts of dogged journalists. Media pressure gets results, as evidenced in this headline: "Google Seeks to Break Vicious Cycle of Online Slander: In Response to *Times* Articles, the Search Giant Is Changing Its Algorithm, Part of a Major Shift in How Google Polices Harmful Content."[55]

A former Google engineer has pointed out that such pressure "can also affect hiring and retention. If Google is seen by engineers who have many job prospects as a place that's doing uncool or unethical work, people will simply take another job elsewhere. It'll be harder for Google to get talent and in some cases to retain the existing talent because people object to these projects."[56]

Sadly, journalism is under great strain because digital entrepreneurs have undermined its economic models. That's why we need to subscribe to and donate to the newspapers, magazines, websites, blogs, newsletters, and TV and radio stations that are doing this crucial work.

Consumers are not the only ones who have a stake in the future of the news media. Maëlle Gavet notes how much content journalists provide—with little or no compensation—to social media platforms, and she suggests that the companies should give these sources syndication rates, as well as a fairer cut of advertising revenues. She also argues that the platforms could do more to support journalism directly through endowments and other measures. In 2019, Facebook announced that it would do so, as part of a major Journalism Project, a plan to spend $300 million dollars over three years to support local journalists and newsrooms.[57]

In 2022, though, the company eliminated much of this funding. Does that sound like a familiar pattern? (Facebook creates some noble initiative such as the Responsible Innovation team, garners favorable press to counter criticism of its past misdeeds . . . and then reduces or ends the program.)

We Can Relinquish or Reject Especially Dangerous Technologies

Daniel Deudney explains that, "In general terms, restraints fall into three broad categories: *reversal* (undoing something that has been done), *regulation* (continuing to do something in circumscribed ways), and *relinquishment* (completely forgoing a technology or activity)."[58]

In regard to the third option, he adds that, "Prudent choice avoids embarking on enterprises that will turn out poorly unless very difficult things are done. The roads to perdition are often paved with baselessly optimistic assumptions about capacities to accomplish and sustain strenuous or heroic arrangements."[59]

Roger McNamee points to many ways in which Big Tech companies are trying to diversify by entering new fields, in ways that would give them even broader power over our lives than they currently have. He suggests that "forcing Google, Facebook, Microsoft, and Amazon to divest or abandon new initiatives such as smart cities,[f] reserve currencies, transportation, financial services, and smart devices would serve the public interest by limiting

---

[f] "New cities" are often designed with built-in technologies that allow a high level of electronic surveillance and gathering of user data, ostensibly with the goal of allowing them to be run more efficiently.

the scope of established monopolies and preventing them from taking control of services normally provided by government."[60]

When I spoke with Deudney, he advocated for a major reversal—"The most important thing we should do is to undo one of our largest and unacknowledged space programs, which is the ballistic missile transportation system"—and several forgoings. Aside from giving up space colonization in general, he said that he would, for the foreseeable future, "relinquish the construction of very large orbital infrastructures, which I believe would have the capacity to dominate the planet." When asked what other technologies he would suggest rejecting for now, he said that "artificial superintelligence would be at the top of my list."[61]

Though out-and-out rejection of just about any new technology will no doubt strike extreme techno-optimists and libertarians as radical, unacceptable, and also impossible, some present such extreme risks that it may be the best, sanest option. As Carl Sagan put it back in 1994, "It's no use saying that all technologies can be used for good or ill . . . When the 'ill' achieves a sufficiently apocalyptic scale, we may have to set limits on which technologies may be developed."[62] Even if a technology may not have existential consequences, some—like autonomous lethal weapons—seem simply *mala in se*, bad in-and-of themselves.

For those who would still argue that technologies, once envisioned, cannot be stopped, it's worth pointing out again that we have already done so in a small but significant number of cases, such as our international bans on chemical and biological weapons, and the Outer Space Treaty of 1967, which forbids the testing and deployment of orbitally based nuclear weapons.[63] We have also seen recent moratoriums on funding for gain-of-function biological studies that make pathogens more potent or contagious (although the U.S. National Institutes of Health lifted its "pause" on those studies in 2023).[64] For the past quarter of a century, in 29 countries, the Council of Europe's Convention on Human Rights and Biomedicine has banned heritable human genome editing.[65] And many countries have banned the creation of cloned human embryos.[66]

Currently, we have an interesting case in which there are two opposing sides to an argument to ban a technology, though both sides ostensibly share the same goal. I'm talking about solar geoengineering to fight global warming. This involves large-scale schemes such as setting up shields in near-space, or dispersing particles into the atmosphere to block some of the sun's rays. Proponents say that we ought to take such actions to counter the imminent threat,[g] while critics say that these actions could have the unintended effect of removing incentives to cut greenhouse gases.[h] Also, they argue that unintended consequences might include affecting the ability of Earth's plants to photosynthesize, or creating unevenly distributed negative impacts, such as droughts, that might cause neighboring countries to go to war.[67]

In 2010, 193 nations agreed on a moratorium on geoengineering under a United Nations treaty called the Convention on Biological Diversity, though it was not legally binding, and still allowed for small experimental projects.[68] In 2022, hundreds of scientists from around the world signed a letter arguing for an absolute moratorium on the technology.[69] Mexico actually instituted such a ban in 2023, after the founders of a U.S. tech startup called Make Sunsets took it upon themselves to send two balloons up into the stratosphere over Baja, California and release particles of sulfur dioxide.[70]

---

[g] Fossil-fuel companies seem to be supporting the idea of geoengineering. Critics argue that this is because it could enable them to resist having to change their ways.
[h] And those gases do things like acidify our oceans, which geoengineering would not prevent.

Of the four current biggest threats to the future of humanity, three are already in existence, and have been for some time. I'm referring to global warming, which was spurred by the first Industrial Revolution; nuclear weapons, which the U.S. military first built at the end of World War Two; and biotechnology, which has made great strides in the last several decades, with scientists devising tools that have made it possible to create extremely dangerous and contagious pathogens.

The fourth threat, artificial superintelligence, has not yet been realized. If all the researchers around the world currently trying to develop it simply stopped, as A.I. researcher Eliezer Yudkowsky has advocated, it never *would* be realized.[71]

In March of 2023, tens of thousands of scientists and tech professionals signed a letter published by the Future of Life Institute, a nonprofit organization devoted to reducing those four risks, which said that "we call on all AI labs to immediately pause for at least 6 months the training of AI systems more powerful than GPT-4. This pause should be public and verifiable, and include all key actors."[72] The letter was signed by such notables as Elon Musk, Steve Wozniak, and a number of A.I. pioneers. It urged that, "AI labs and independent experts should use this pause to jointly develop and implement a set of shared safety protocols for advanced AI design and development that are rigorously audited and overseen by independent outside experts. These protocols should ensure that systems adhering to them are safe beyond a reasonable doubt."

Since specific, limited uses of A.I. have been beneficial, especially in realms such as science, mathematics, and medicine, the authors of the letter made it clear that, "This does *not* mean a pause on AI development in general, merely a stepping back from the dangerous race to ever-larger unpredictable black-box[i] models with emergent capabilities."

The researchers at the Institute came up with a bunch of excellent recommendations:

> "AI developers must work with policymakers to dramatically accelerate development of robust AI governance systems. These should at a minimum include: new and capable regulatory authorities dedicated to AI; oversight and tracking of highly capable AI systems and large pools of computational capability; provenance and watermarking systems to help distinguish real from synthetic and to track model leaks; a robust auditing and certification ecosystem; liability for AI-caused harm; robust public funding for technical AI safety research; and well-resourced institutions for coping with the dramatic economic and political disruptions (especially to democracy) that AI will cause."

The letter was eminently sane, intelligent, and reasonable—but there were two big problems with it. First, although it did a great job of bringing the dangers of A.I. to public attention, it wasn't clear what a six-month pause could actually accomplish. Eliezer Yudkowsky has done a great deal of research into A.I. safety, and as he has taken pains to point out, we're nowhere near having the capability to devise controls that would make advanced models safe beyond a reasonable doubt, and a few months of research wouldn't bring us much closer.[73]

Second, despite the signatories' best intentions, the pause *didn't even happen.* Six months later, in a podcast titled "All Gas,

---

[i] This refers to the fact that researchers can give their models a prompt, and see what comes out on the other end, but they often can't trace what the models do in between. A.I. provides a very odd exception to the history of technology: surely there has never been another field in which the workings of its machines could be described as "inscrutable" and "incomprehensible."

No Brakes in A.I.," tech journalists Casey Newton and Kevin Roose noted that no A.I. companies had instituted a hiatus in their development of the technology, and it actually proceeded at a *greater* rate.[74][j]

This is just one of many examples of why we urgently need to break the power of the Wheel.

---

[j]   Though Musk was one of the signers calling for a pause, just four months after the release of the letter he announced the launch of his own A.I. company, xAI. (https://www.reuters.com/technology/elon-musks-ai-firm-xai-launches-website-2023-07-12/)

Chapter 18
Other Ways To Go

"Power and social control, once exercised chiefly by military groups who had conquered and seized the land, have gone since the seventeenth century to those who have organized and controlled and owned the machine. The machine has been valued because—it increased the employment of machines. And such employment was the source of profits, power, and wealth to the new ruling classes."

—Lewis Mumford[1]

"It is easier to imagine the end of the world than to imagine the end of capitalism."

—Professor Fredric Jameson[2]

IT WOULD BE FOOLISH TO IGNORE THAT THE RE-ELECTION OF Donald Trump marked a massive setback for those concerned about improving Big Tech (or reversing global warming, or preventing pollution, or strengthening democracy). If we want to be clear-eyed, we'll have to sadly acknowledge that—in the U.S. at any rate—a number of these objectives will have to await the election of a more sympathetic administration.

I have to concede something else. In the last couple of chapters, we looked at a number of actions that could avert, ameliorate, or reverse negative impacts of technology, especially digital tech. If we can take such measures, we'll have achieved a great deal, and set ourselves on a path that might lead toward a brighter future. Yet we still might not have solved our biggest problems.

As Shoshana Zuboff puts it, "We can't rid ourselves of later-stage social harms unless we outlaw their foundational economic causes. This means we move beyond the current focus on downstream issues such as content moderation and policing illegal content. Such 'remedies' only treat the symptoms without challenging the illegitimacy of the human data extraction that funds private control over society's information spaces. Similarly, structural solutions like 'breaking up' the tech giants may be valuable in some cases, but they will not affect the underlying economic operations of surveillance capitalism."[3]

Dipayan Ghosh even argues that, "Consumer internet executives will secretly encourage public debates over fake news and hate speech. They will add fuel to these flaring deliberations for as long as they can, drawing our eyes away from the subtler, more fundamental problems at the heart of the industry's commercial regime."[4]

In a larger sense, the problem may not just be "surveillance capitalism," or one industry's financial models, but the fundamental economic system that drives them. And that, of course, is our modern, hyper, unrestrained version of capitalism.

As journalist Ben Tarnoff puts it, in a statement that could apply to many forms of tech, "The profit system produces the dysfunctions and depredations of the modern internet. Today's internet reformers would leave this system intact. [...] Even with the best regulatory and antimonopoly measures, corporations would still own the internet. Immensely consequential decisions would be left in the hands of executives and investors. Most people would have no say in matters that centrally affect their lives."[5] The forces of this system are so strong in today's tech world that it has become extremely difficult to challenge them.

In an earlier chapter we looked at the phenomenon of *lock-in*. When I spoke with L.M. Sacasas, he said that,

"As time passes, and especially as systems coalescence around a technology, so for the automobile, for example, or the power grid, then there's a kind of momentum that sets in where it becomes almost prohibitive to go back and change things or set out on a new path. We've known this for a while with regards to some of the changes that we might want to see deployed for the sake of changing the current trajectory of climate change. You find that you have this global system of automobile manufacturing and fossil fuels that has financial vested interests, of course, but also an infrastructure that we can't simply—from one day to the next or one year to the next—overhaul and transform. And so things feel as if they're inevitable. But if you go back to the beginning of the 20th century, there would have been a lot more flexibility with regards to the direction of these paths that could have unfolded."

It seems clear that this tendency toward lock-in has applied not only to many of our technologies, but to our chief economic system. As Kim Stanley Robinson put it:

"Economics as it's practiced now is the study of capitalism. It takes the axioms of capitalism as givens and then tries to work from those to various ameliorations and tweaks to the system that would make for a better capitalism, but they don't question the fundamental axioms: *everybody's in it for themselves, everybody pursues their own self-interest, which will produce the best possible outcomes for everybody.* These axioms are highly questionable, and they come out of the eighteenth century or are even older, and they don't match with modern social science or history itself in terms of how we behave, and they don't value the natural biosphere properly, and they tend to encourage short-term extractive gain and short-term interests. These are philosophical positions that are expressed as though they are fixed or are nature itself, when in reality they are made by culture."[6]

Author and technical book publisher Tim O'Reilly says that, "If you think back to the Middle Ages, everybody believed in the divine right of kings. Right now, everybody believes in the divine right of capital. It's only natural that the owners of businesses and the owners of capital should try to extract as much as possible for themselves and leave society in the lurch, and that they should basically treat people as a cost to be eliminated. We accept that. When I say 'we,' I mean all of us accept it.[a] We simply believe that that's the way the world works."[7]

O'Reilly argues that "in some ways financial markets are that rogue AI that people like Elon Musk have been talking about." He says we're now in "a world in which you have an increasingly algorithmic financial system saying, 'Hey, optimize for corporate profit because it drives stock price. Never mind what happens to the people. Never mind what happens to society'."

I think I ought to take a moment here and state something obvious: that there's nothing inherently wrong with using a technology to conduct business, or to promote that business through advertising.[b] As Sacasas has put it, "The point is not that

---

[a]   Of course, some of us don't, but he's being extreme to make a good point.
[b]   As Douglas Rushkoff made sure to point out to me, "Not all advertising is bad.

commerce is bad. I buy things. You buy things. We all need to buy things. I'm glad to pay someone for their labor. I'm glad to be paid for mine." But he added that, "The point, and certainly not an original one, is that we should be wary of allowing the logic of the market to colonize all facets of our experience. [...] Aspects of the world now appear to us framed by the implicit challenge: Commercialize this."[8]

In some respects, capitalism can certainly be a force for good, contributing to rising standards of living in developing countries, providing money for research and innovation, and enabling the success of companies that make useful products. Events of the past couple decades, however—ranging from a staggering rise in global economic inequality[9] to the massive financial crisis of 2008—have shaken even some of the system's most enthusiastic cheerleaders. Too often, it tends to prioritize the profits of a minority of shareholders over the safety, health, and sanity of a majority of stakeholders (i.e., everyone else.)

Computer programmer Nick Drozd took aim at this Wheel factor when he reflected on Nick Bostrom's famous thought experiment: "Suppose that instead of maximizing paperclips at all costs (which is obviously a farcical example), a runaway AI had the goal of maximizing shareholder value at all costs. But that's exactly the goal of corporations today. So corporations have a huge incentive in seeing that shareholder-value-maximizing AIs are created."[10]

In other words, maybe what we should be worrying about most right now is not some future Skynet or HAL-9000[c]—it's the Silicon Valley founders and executives who are desperate to dominate all sorts of fields of tech.

Marketing and advertising is the way you tell people about something that exists. Just like not all drugs are fentanyl patches in the alley of the 7-11. There's also Advil and antibiotics."

[c] The rogue artificial superintelligences in *The Terminator* and *2001: A Space Odyssey.*

Critics have offered a great deal of speculation about whether it would be possible to shut down an artificial superintelligence gone rogue.[d] But as Ted Chiang puts it, "*Capitalism* is the machine that will do whatever it takes to prevent us from turning it off."[11]

When it comes to the potential for human job loss, all the talk about the amazing benefits of artificial intelligence can be something of a red herring. As Chiang puts it, "As it is currently deployed, A.I. often amounts to an effort to analyze a task that human beings perform and figure out a way to replace the human being. Coincidentally, this is exactly the type of problem that management wants solved. As a result, A.I. assists capital at the expense of labor."[12] In other words, the question is not *Will A.I. put humans out of work?* but *Will human employers use the technology to eliminate the cost of hiring human workers?*

For employers, the benefits of robots and A.I. systems are obvious: they can work 24 hours a day, seven days a week; they never try to organize unions; and they don't expect any wages. They may not even need *lighting* in which to work—a company called Symbotic[13] has even sold automated, multi-million-dollar warehouse systems that can operate almost totally in the dark. That's because—beyond a small handful of managers—there are no human workers inside.

Proponents of A.I. love to talk about how the technology will free humans from having to work; supposedly, we'll all live enriching lives of leisure and creativity and play. But this raises an

---

[d] President Obama joked about the emergence of artificial superintelligence that, "You just have to have somebody close to the power cord. Right when you see it about to happen, you gotta yank that electricity out of the wall, man." (https://www.wired.com/2016/10/president-obama-mit-joi-ito-interview)

obvious question: what, exactly, is the mechanism by which the profit a tech billionaire makes from machine-based job replacement will be shared with the workers they have put out of their jobs?

Short answer: there isn't one.[e]

A century ago, Lewis Mumford considered the notion of human workers becoming replaced by machines, and he foresaw a possible outcome: "Lacking the power to buy the necessities of life for themselves, the plight of the displaced workers reacts upon those who remain at work: presently the whole structure collapses, and even financiers and enterprisers and managers are sucked into the whirlpool their own cupidity, short-sightedness and folly have created. All this is a commonplace: but it rises, not as a result of some obscure uncontrollable law, like the existence of spots on the sun, but as the outcome of our failure to take advantage by adequate social provision of the new processes of mechanized production."[14,f]

Seeing such a collapse as ever more possible, some tech moguls envision being targeted by hordes of angry, pitchfork-wielding citizens who may not only be unemployed, but *unemployable*. If that happens, Peter Thiel knows what he'll do: escape to the 477-acre

---

[e] Some may claim that these tycoons will make up for the loss of jobs by paying massive income taxes and thus ensuring a social safety net, but this argument runs into an obvious problem: they're notoriously good at evading taxes. A *ProPublica* investigative report noted that Jeff Bezos didn't pay *any* federal income taxes in 2007 and 2011; in 2018, Elon Musk paid none either. (https://www.propublica. org/article/the-secret-irs-files-trove-of-never-before-seen-records-reveal-how-the-wealthiest-avoid-income-tax)

[f] What's more, the economic effects of job replacement might go far beyond the impact on the labor force. Humans pay taxes on their earnings; A.I.s don't. This could reduce our ability to pay for education, infrastructure maintenance, policing, Medicare, Social Security, and all of the other functions that our taxes normally support. Aside from that, human workers spend a large part of their earnings on clothing, food, manufactured goods, services, etc. A.I. models don't need these things, and that might dry up these normal economic flows.

property he has prepped as a refuge in rural New Zealand.[15] In 2023, *Wired* magazine reported that Mark Zuckerberg was building a massive $270 million compound in Hawaii equipped with a 5,000-square-foot underground shelter "and what appears to be a blast-resistant door."[16]

Other tech titans have thought about ways to forestall a rebellion of the unemployables, and have begun to talk about establishing a guaranteed Universal Basic Income for everyone. I saw one Facebook commenter greet this with considerable skepticism: "The same people that put in the robots, so they would not have to pay us to work anymore, will be paying us a living salary for not working. And a free unicorn pony for every child."

This might be a good time to ask a few essential questions about labor and technology and economics.

What are companies ultimately <u>for</u>? Are they means of providing human beings with the opportunity to make a living, create useful products, and provide helpful services—or are they just vehicles for generating profits for their owners and shareholders?

What are our technologies for? Are they for making our working lives better, or for putting many of us out of work?

Should our end goal be greater efficiency, growth, and profits for corporations, or greater health, security, and happiness for all?

As we have seen, some tech leaders and investors have an overpowering tendency to prioritize profit over people, and the result could be catastrophe.

What can we do about it?

I see two main schools of thought.

Proponents of the first argue that we can work within the system, and reform and restructure it so that it will better serve everyone. In an uncharacteristically upbeat moment, Shoshana Zuboff said that she believes that once "the sleeping giant of democracy awakens to actually confront [the problems with digital technologies], we're creating space for new competitive solutions. Because we need different kinds of companies, and different kinds of capitalists, and different kinds of ecosystems and alliances that are actually capable of reclaiming the digital for the kind of values and functionality that we wanted from it in the first place."[17]

We might dial back the most aggressive, invasive forms of capitalism. Many tech and media companies make most of their earnings through advertising, but we have other financial models—such as subscriptions—that could provide different forms of revenue. As Roger McNamee puts it, "the best way to differentiate the good from the bad is to look at economic incentives. Companies that sell you a physical product or a subscription are far less likely to abuse your trust than a company with a free product that depends on monopolizing your attention."[18] Personally, I'm happy to subscribe to media sites that provide me with good journalism, and to pay for streaming TV services, rather than sit through endless ads.[g]

Jaron Lanier observes that "when people started paying for Netflix, we got what we call Peak TV. Things got much better as a result of it being monetized to subscribers. So if we outlawed advertising and asked people to pay directly for things like Facebook, the only customer would be the user. There would no longer be third parties paying to influence you. [...] We might actually see things improve a great deal."[19]

Of course, the subscription model only works if you've got the money to pay for it; otherwise, it's hard to ignore the appeal of

---

[g] Since I was a kid, the number of minutes of commercials in one hour of broadcast TV has approximately *doubled*.

"free" content. But even quite low rates might make the change-over feasible: if Facebook users only paid fifty cents a month to use the platform, for example, that would still add up to billions of dollars per year in revenue.

In theory, venture capital, the economic model that now powers Silicon Valley, seems to make good sense. You take young people who might have great ideas for products or services but little money or business sense, and pair them with older mentors who have lots of both.

Too often, though, V.C. firms operate in a style of capitalism jacked up on steroids. Whereas traditional investors might be concerned about ensuring that a company is soundly run so it can continue to earn revenue far into the future, V.C.s are sometimes just concerned with how much they can pump up the value of a company in the near term. This is not just a problem with the V.C. model, but the ways in which our legal and tax systems encourage such shortsighted behavior. As Douglas Rushkoff told me, "Taxes on distributions on revenue are very, very high, whereas taxes on capital gains, which means selling the whole company, are very, very low. So our whole tax code encourages making companies that just grow exponentially."[20]

If a founder starts out with an idealistic vision, it may well get corrupted along the way. "If you have a good idea," says Rushkoff, "and then bring it to Goldman or Morgan Stanley in order to get an IPO, now you're doing something else, right? It's not about just keeping your little idea alive. It's about—whether you know it or not—using your idea to generate exponential, unicorn-like returns. And where does that come from? That comes from exploiting and extracting. So it's sort of necessary to go negative at that point."[21]

As we've seen in earlier chapters, venture capitalists keep

showing up as agents of the Wheel, but again, there's nothing inherently wrong with investors supporting fledgling tech companies, especially if they consciously work to have a more beneficial impact on society. One example is Kapor Capital, founded by Freada Kapor Klein, who had a background in nonprofit work, and her husband Mitch Kapor, who made a lot of money in spreadsheet software. They started a firm that explicitly prioritizes backing entrepreneurs in low-income communities and communities of color.[22] They have invested in more than 170 startups with a socially positive angle in fields such as financial service and health care, including a company bringing low-cost tutoring to students in low-income neighborhoods, and another helping people in those areas pay their bills and improve their credit scores. Lest this sound like a purely philanthropic venture, Klein and Kapor talk about "sustainable capitalism," and argue that "doing well and doing good are not fundamentally opposed."[23] They say that Kapor Capital's financial results put it in the top quartile of funds of comparable size.

"When we look at a business," Mitch has said, "one of the important questions we ask is, 'If this business succeeds, who is going to be better off, and who is going to be worse off? Who is it serving?'"[24]

When it comes to how to run companies, Douglas Rushkoff has spent much of his career thinking about the intersection of technology and finance, and he also recommends a different way to operate. In fact, it's not so different at all—part of what he advocates for is a return to a more traditional notion of what a business should be. "The whole thing you do," he told me, "is just position it so that you are building a company that's going to earn money for its investors through revenue, rather than selling the company itself."[25] He says that the solution involves "rejecting the notion that the only healthy career, company, or economy is one that grows at a rate defined arbitrarily by a bank, a group of investors, or the startup ethos now so dominant in Internet culture,"[26] and he argues for using models that are "less dependent on establishing and enforcing monopolies and less encumbered by the growth imperative."[27]

In other words, the executives of a sound, sane company, who want it to be viable for the long run, shouldn't need to keep growing it at a frenzied rate in order to keep it alive, and they shouldn't have to gobble up all of its competitors. They should just need to provide steady earnings, and enough profit to benefit the founders, workers, and investors.

Rushkoff puts the idea in a historical context, pointing out that —back in the 1800s—economic philosopher John Stuart Mill "saw the end of growth concluding in what he called the 'stationary state'—a sustainable equilibrium in which there would be 'a well-paid and affluent body of labourers; no enormous fortunes, except what were earned and accumulated during a single lifetime; but a much larger body of persons than at present, not only exempt from the coarser toils, but with sufficient leisure, both physical and mental, from mechanical details, to cultivate freely the graces of life'."[28]

This is not just a matter of designing more benevolent technologies—it's a matter of designing *companies* that are oriented toward developing them, and not incentivized by priorities that might lead to harms. The idea is not to rely on the beneficence of top executives to make this happen; there are ways to write positive objectives into the very structure of a company. For example, while the ratio of CEO pay to worker pay has risen at a shocking rate in the past few decades, companies can cap CEO compensation at a much more reasonable ratio.[h]

---

[h] In 1984, management consultant Peter F. Drucker argued that a CEO should not be able to make more than 20 times the pay of front-line employees. (https://www.sec.gov/comments/df-title-ix/executive-compensation/executivecompensa

Surprising as it may sound, some companies even institute voluntarily restrictions on their profits, because they see that this can free them from ravenous external pressures. "For example," notes Rushkoff, "many are toying with the *benefit corporation* as a way of tempering the emphasis on short-term and extractive profit suffered by traditional corporations goosed up on digital systems. A benefit corporation is expected to pursue profits, but that profit motive must be secondary to a stated social or environmental mission. By law, share price must take a backseat to something else, something decidedly beneficial."[29]

OpenAI was originally designed as a nonprofit, with a cap on employee and investor profits. As CEO Sam Altman put it, "One of the incentives that we were very nervous about was the incentive for unlimited profit, where more is always better. And I think you can see ways that's gone wrong with profit, or attention, or usage, or whatever, where if you have these well-meaning people in a room, but they're trying to make a metric go up into the right, some weird stuff can happen. And I think with these very powerful general purpose A.I. systems, in particular, you do not want an incentive to maximize profit indefinitely."[30] The founders also wrote into the company charter that its fiduciary responsibility was to create A.I. that would benefit all of humanity. And they said that the company's board would have the power to shut things down if they felt that safety was not being taken into enough consideration.

"I thought that was an interesting moment in the history of Silicon Valley," Steven Johnson told me. "They're like, 'Oh, here's the new big tech advance that's coming down the pipeline, and it's not safe to let a private company—a normal venture-backed

---

tion-60.pdf). That was the average in the prosperous 1950s. In the U.S. in 2020, however, researchers at the Economic Policy Institute found that the average pay ratio of CEOs to employees was 351 to 1. (https://www.epi.org/publication/ceo-pay-in-2020/)

startup—release this into the world. We can't let that happen. It's too dangerous."[31,i]

Unfortunately, although the founders of OpenAI kept their capped-profit model, they soon decided that they needed major investment so they could compete with companies like Google, so they made plans to take on more than ten billion dollars in backing from Microsoft[32]—and then they rushed to release their chatbots, doing more than anyone else to spur the frantic A.I. arms race.[33] What's more, the 2023 incident in which OpenAI's board fired Sam Altman as CEO, but then was almost immediately forced to reinstate him, showed that when push came to shove—when the needs of shareholders and corporate partners came to the fore—the board was *not* able to rein in the company's leader.[34] After that, the two leaders of its long-term alignment team, including cofounder and chief scientist Ilya Sutskever, left the company, citing safety concerns, and OpenAI dissolved the team.[35,j]

Here's another way to change the current model: companies can write in a requirement that investors must hold onto their shares for substantial lengths of time, so they won't pressure the founders to quickly sell. Rushkoff mentions successful companies such as Meetup and Kickstarter[k] whose founders explicitly designed them so they could resist the growth imperative, and there are a few successful tech nonprofits, such as Wikipedia and the Internet Archive.

---

[i] Johnson qualified his admiration, though: "What they haven't figured out is, how do you decide what's good for all of humanity? If it's just a bunch of people in a boardroom in the Mission District, who are all billionaires, that doesn't seem like we really advanced the process at all."

[j] Shortly thereafter, the company established a new safety committee, led by none other than Sam Altman. (https://edition.cnn.com/2024/05/28/tech/openai-announces-new-safety-board-after-employee-revolt/index.html)

[k] Its founders turned it into a benefit corporation in 2015.

The Frankenstein Fix

I should mention that some people *don't* see restructuring corporations as the chief solution to what goes wrong with technologies. When I asked journalist Noah Kulwin about the idea, he scoffed. "I don't think that there's a way out that you can get through companies." He explained that, "Tech companies are predicated on delivering gigantic returns for their venture capitalists. If you go into a meeting with any serious venture capitalist, who is a good investor and makes money for their limited partners, that person will laugh you out of the fucking room if you were to tell him, 'I actually plan to *not* deliver 100X return, because I focus on sustainable and equitable growth built on a business model that treats all of my suppliers and consumers and workers fairly'."[36]

It would seem, then, that it might be a good idea to create alternative funding sources for startup founders. Kickstarter uses a different model, crowdfunding. And Robert Gardiner suggested sovereign wealth funds. Wendy Liu expanded on that idea, saying that we could:

"create a publicly owned investment fund whose scope is limited to non-profit ventures. This would be similar to what's traditionally considered 'social entrepreneurship,' with socially beneficial aims and no pressure to turn a profit, but in this case there would be ample access to funding. The government could provide the upfront capital and take majority ownership but otherwise get out of the way of daily operations—similar to how most investors work now. Pitch decks and financials would be judged on the basis of practicality and social value rather than market domination or profitability, and fraud or malfeasance would be severely held to account to dissuade Theranos-style grifts. This fund could also provide the necessary capital for existing corporations whose workers would like to convert it into a worker-owned co-op, or another suitable structure enabling worker control of production."[37]

Liu cites the U.K.'s Government Digital Service, created in 2011, which provides small and medium-sized tech companies with public funding for research.

Hypercapitalism is a system that works incredibly well for a minority of people who are shrewd enough to know how to exploit it. (Or were simply born wealthy). But it's not working well for billions of people across the world, many of whom have turned to populist demagogues who can channel their resentment at getting left behind as the Elon Musks and Mark Zuckerbergs become more and more staggeringly rich.

Jaron Lanier has thought for a long time about how to return everyday citizens to the center of our digital economies, arguing that we users should be able to make money by selling access to our own data and content, rather than letting big companies leech that value away from us.

"One fine day," he wrote back in 2010, "your ISP could offer you an option: You could stop paying your monthly access charge in exchange for signing up for the new social contract in which you pay for bits. [...] If you chose to switch, you would have the potential to earn money from your bits—such as photos and music—when they were visited by other people. You'd also pay when you visited the bits of others. The total you paid per month would, on average, initially work out to be similar to what you paid before, because that is what the market would bear. Gradually, more and more people would make the transition, because people are entrepreneurial and would like the chance to try to make money from their bits."[38]

When I read that back then, it seemed that it would be quite difficult, technically-speaking, to enable this system of making many tiny payments for the use of many tiny amounts of data, but some people now envision that the new blockchain technology could provide the medium that would be able to keep track.

In *The Ministry for the Future*, Kim Stanley Robinson conjures up "a whole new internet ecology" featuring a co-op called YourLock, owned by users, in which their data is encrypted and secure, and they "could use it as a negotiable asset in the global data economy, agreeing to sell your data or not to data-mining operations out there who quickly saw the new lay of the land and began to offer people micro-payments for their data, mainly health information, consumption patterns, and finance. The royalties for being oneself in the world machine were not insignificant, a kind of lifetime annuity, small but useful."[39]

As someone who has created a goodly amount of content in my life, I'd be excited about being able to cut out some of the middlemen who have hoovered away much of its value, yet I have to side with a number of critics who are not so enthused about this idea of users monetizing their own data.

As Rushkoff told me, "I understand what Jaron wants. He feels that if people are generating data as they move through their lives, they should be able to be paid for it—certainly, if the data's being used by someone else. I understand his impulse to want to get more human behavior on the books so that we can be compensated for it, but I kind of have the opposite view. I want to take more human behavior *off* the books, so that we live in more of a favor universe than one in which a blockchain is measuring each piece of value that everyone's creating for everybody else all the time. So yeah: it's technologically feasible, but I just don't know if I like the world that it creates."[40]

Maëlle Gavet warns that if we required digital companies to buy our data from us, "at that point people would lose all control over how the platforms used it."[41] And L.M. Sacasas told me that the notion "amounts in many respects to just the increasing colonization of economic rationality, so that we are now able to commodify more and more of our experience, or turn more and more of experience into economic resources. [...] I'm grateful for a lot of Lanier's work and his insights, but if the solution is simply to make sure we're getting our fair share in economic terms, I don't know if it misses the point so much as it's a solution that creates another set of, or feeds into another set of problems."[42]

Some predict that self-monetization of user data could take a distinctly unhappy turn, as explained in this exchange between journalist Ezra Klein and writer Dan Olson:

Klein: "Take TikTok as it exists now and how alluring and seductive it is, and now imagine that every time you're some teen whose TikTok gets 50,000, 150,000, a million hits completely randomly, you actually get a bunch of money. Not a ton of money, but enough that it matters to you. How much more addictive does TikTok become to you when it really is a slot machine that occasionally pays out coins? And is that a better world?"

Olson: "It does strike me as horrifying, the potential for that. I've been spending a lot of time looking at other content communities online that are heavily industrialized in this sense, like content mills, and really asking myself, what is it like to work in a content mill? So you have all these content mills that churn out endless reams of garbage about you know, acai berries and blueberries and how oatmeal will help you grow your hair back or whatever. And humans are writing that. Someone somewhere, a creative person or a skilled writer, or at least, a like, aspiring writer, is sitting somewhere in a room, using their fleshy human hands to type on a keyboard and will those words into existence.[1] And that incentive cage is basically what you're describing there. It's just taking the incentives and structure of a content mill and encaging, enclosing, if you will, the entire internet in it."[43]

Here's another idea about how to work within the system.

In *The Ministry of the Future*, Robinson devotes a lot of energy to coming up with realistic ways to change our economies before global warming is irreversible. Recognizing that a worldwide anti-capitalist revolution is highly unlikely to happen within the next several decades, he imagines how we might perform a kind of financial ju-jitsu, using capitalist incentives to drive our current system toward healthier goals. First, the protagonists at the Ministry convince representatives of the world's most powerful banks that there's no future for their institutions—and for the stability of money itself—if global warming gets much worse. Building on today's current carbon taxes and cap-and-trade systems, they come up with a system of "carbon coin currency" in which corporations can earn "one coin per ton of carbon-dioxide-equivalent sequestered from the atmosphere, either by not burning what would have been burned in the ordinary course of things, or by pulling it back out of the air."[44] Over the course of some years, the protagonists see this strategy begin to have a significant impact in reversing climate change: "The majority [of carbon] was being drawn down by reforestation, biochar, agroforestry, kelp bed and other seaweed growth, regenerative agriculture, reduced and improved ranching, direct CO_2 capture from the air, and so on.

---

[1] Of course, nowadays a human may *not* be writing that, and not getting paid for it—an A.I. model might be writing it for free.

All these efforts were paid for, or rather rewarded beyond the expense of doing them, in carbon coins."[45]

As with the user monetization of data, though, not everyone loves the idea. I asked Rushkoff what he thought of it, and he replied, "So what you're asking is, can we gamify capitalism in such a way as to incentivize behaviors that fix things? Maybe. Usually not, because once you have a single token working towards a single outcome, you end up externalizing harm to other outcomes."

He offered an example: "When the Ministry of Health in England followed Maggie Thatcher's incentivized program for waste reduction in hospitals, they came up with the idea that what they would do was give funding to hospitals based on how quickly they could get patients from over-crowded emergency rooms into wards. What clever hospitals decided to do was to rename their corridors 'wards,' and have gurneys that they could take the wheels off of. So a person's on the gurney, waiting for treatment. You wheel them to the hallway and take off the wheels. Then you declare them in a ward, rather than waiting. And that worked, and those places got funding. But the actual waiting time increased."

Rushkoff warned that "mechanical, over-simplified, numeric and digital solutions lead to unintended consequences, even from those who may have everyone's best interest at heart, if not fully in mind."[46] "I'm generally hesitant to use metrics to motivate better behavior," he concluded. "I prefer people to actually be in the bigger game of solving the problem rather than satisfying an incentivized metric."

I was reminded of the economic term *perverse incentive* when I read a recent op-ed by two pioneers of carbon capture technology. Their original goal was to benefit the environment and slow global warming by making it possible to grab onto carbon dioxide emissions before they could leave power plants, but they now warned that this can have the unintended effect of "slowing the transition away from fossil fuels."[47] They said that when governments give companies credits for using the technology, they're rewarding the ones who produce the extra CO_2 in the first place, rather than incentivizing them to turn to cleaner types of energy.

Despite such drawbacks, I think we might still find better ways to incentivize companies to move in more positive directions. But we can't just rely on them to do so: we still need regulations—and effective penalties—for those who cause harms.

I realize it may seem that I keep mentioning people who have come up with possible helpful solutions, only to offer other voices that shoot those solutions down, but Steven Johnson puts a positive spin on this kind of dialectic. He told me that, "We should remind ourselves that with climate change and superintelligence and AI, we're having debates now that are actively wrestling with what we think might happen in thirty years or fifty years. And, that's a new thing. People didn't think that way before. Nobody inventing the steam engine was like, 'Yeah this is really helpful and great. But I think potentially we're outputting this stuff into the atmosphere with all this carbon that could be a problem in a couple centuries'."

"People built things to last," he noted, "but they very rarely built things in a way that was mindful of how they could be abused, or secondary effects that they could produce on the scale of generations in the future. And now we're trying to think that way. And the fact that we're even trying is a sign of progress. It's great. But we're new at it. And so that's why I feel like there's just a lot of different mechanisms out there that we haven't tried yet."

If we want to reach the best solutions, the first step is to make ourselves aware of current problems with our technologies and our economies. The second is to figure out what's causing those problems. The third is to ask how we might structure things differently. As Wendy Liu put it, "Could there be another way to do it? Could there be a way to develop technology *without* the role of a multinational corporation necessitating patents and lawyers and stock price analysts? Could there be a way to build software that people wanted, without having to resort to business models that subjected users to endless distracting advertisements they would rather not see? Could there be a way to develop technology as a public good, shorn of the idea that it must make money in order for investors to recoup their investment?"[48]

Once we start thinking that way, all sorts of other possibilities emerge. Some of them involve working *outside* the usual system.

While laissez-faire capitalism has allowed a few individuals and corporations to seize the lion's share of the world's property, and socialism and communism put the rest under the control of the state, Rushkoff advocates for a third way of doing things, called *distributism*. This philosophy calls for spreading ownership as widely as possible—and spreading earnings and prosperity as well.

Steady sustainability of businesses is the priority, rather than endless growth. Instead of value being extracted upward (and often *out of*) companies, it's ploughed back into them, used to invest in infrastructure and reward the workers who enable their success. I use the verb *ploughed* intentionally, since—as Rushkoff points out—this philosophy reflects a return to values of community and family and cooperation that governed economic life in medieval villages and towns in the centuries before voracious capitalism took over. That was a time when peasants earned their livelihood by making, growing, and trading things with each other. Rushkoff notes that the members of a small village might have given a new blacksmith a barn and meals to support him while he got his business up and running, because they knew that his work would eventually help their community, and he notes that crowdfunding could be a modern-day equivalent. He believes that today's digital technologies can provide new opportunities for the kinds of sharing and cooperation that the distributist philosophy promotes.

If this all sounds like something cooked up by leftist revolutionaries, it might amuse you to know that its original ideas were set out by two Catholic popes (Leo XIII, in 1891, and Pius X, in 1931).[49] More recently, Pope Francis was also a proponent,[50] angering capitalists around the world, many of whom fear having their profits wrested away. But Rushkoff explains that distributism is "calling not for the *re*distribution of earnings or capital through taxes or state action after the fact, but for the widest possible distribution of the means of production as *pre*conditions for a healthy marketplace. Workers ought to own the tools they use, and their contributions to an enterprise should earn them an ownership stake in the business itself."[51]

The contrast with today's massive global corporations is stark. As Rushkoff puts it, "While the unimaginative big businesses of today see in the Internet a new way to automate labor, devalue human contributions, securitize wealth, build platform monopolies, and stage spectacular exits, stakeholders in tomorrow's economy should be able to see an opportunity to participate in self-sustaining, highly reciprocal, peer-to-peer, worker-owned, and community-defined marketplaces. Instead of simply digitizing industrial extraction in the name of growing more capital, our new media technologies can *distribute* value creation in the name of a sustainable economy."[52]

Distributism is not just an abstract, untested theory. Kim Stanley Robinson is fond of mentioning a real working example called the Mondragon Corporation. This federation of 95 worker cooperatives, based in a river valley in the Basque region of Spain, was founded in 1956 by a Catholic priest named José Maria Arizmendiarrieta, and it's still in operation today. Its co-ops do things like manufacture industrial machines, bicycles, and refrigeration equipment, and they also operate in the sectors of finance, retail, education, and research. A majority of workers are co-owners of their businesses, and they enjoy good benefits, including parental and sick leaves, medical insurance, and generous pensions.[53]

"The profits don't get shifted out as shares to shareholders," Robinson explains, "but are rather divided three ways, with a third distributed among the employee-owners, a third devoted to capital improvements, and a third given to charities chosen by the employees."[54] The wage ratio between management's top salary and the minimum level of pay is set at no more than six to one.[55] The federation behaves differently from most corporations in other ways too. When times turn bad due to a recession or other economic downturn, companies tend to lay off workers in droves. But the co-owners of Mondragon co-ops find ways to help each other and keep workers working, sometimes by sharing funds, or voting to temporarily reduce their own salaries, or reassigning members to other more viable jobs.

"It's easy to assume that such arrangements must impair productivity," writes journalist Nick Romeo. "But multiple academic studies have found that cooperatives with worker governance and ownership are as profitable as or more profitable than ordinary firms."[56]

This is not some tiny, quirky experiment: as of 2022, the Mondragon Corporation was the world's largest worker co-op. It employed more than 80,000 people, was the 10th largest corporation in Spain, owned a number of international subsidiaries, and in 2023 had annual revenues of more than 11 billion euros.

Mondragon is hardly the only business entity that is centered on its workers, rather than its top executives. Tim O'Reilly points out that, "There are these isolated companies that are playing by totally different rules. A co-op like REI[m] is a good example. REI outperforms all of their public market competitors in measures of traditional, 'real' market activity: Their same-store sales growth is higher, their revenue per store is higher, and they pay more to their employees." He gives another example: "The Green Bay Packers, a storied football team, are owned by the fans. The fans use their ownership to keep season ticket prices low. They're not any less successful than other companies. In fact, they may be more successful."[57]

Notions of cooperative structures are also reaching into the tech world. The anonymous startup founder cited by Tarnoff and Weigel notes that "there are ideas floating around like platform cooperatives, where platforms are owned and governed by a combination of their creators and their users." He mentions that his "favorite video game studio, Motion Twin, describes itself as an 'anarcho-syndicalist workers' cooperative,' and they have a close relationship with their community of gamers." He concludes that "these alternative structures are possible in tech," and "we should be trying out lots of things and seeing what works."[58]

Unfortunately, workers who organize cooperatives must buck a trend toward setting power even more strongly within corporate control. As Wendy Liu explains it, "One major factor underlying the success of many tech companies is the background phenomenon of privatization in many parts of the world. Accordingly, a whole landscape of sectors which *could* have been public services are instead easy prey for tech companies: healthcare, education, banking, mobility, community, and housing, to name a few. To reverse this, we need to excise capital and restore public ownership."[59]

Ben Tarnoff makes a similar argument. "To build a better

---

[m] Recreational Equipment Incorporated, which sells gear for outdoors and fitness activities.

internet, we need to change how it is owned and organized—not with an eye toward making markets work better, but toward making them matter less. Deprivatization aims at creating an internet where people, and not profit, rule."[60]

He offers examples of what it might be like to live in a deprivatized system: "As you click the links in your feed and are transported to other corners of the Web, you can be confident that your privacy is secure. That's because the rights to your personal data are held by a cooperatively owned data trust. You and the other members get to decide under what conditions an online service has access to your data, and under what conditions more data can be created. For instance, your trust might choose to ban the sort of sweeping surveillance that is so integral to online advertising." And, "Maybe you don't live near public transit, so you use an app to call a shared ride. The service is a cooperative, owned by its workers. Unlike Uber and Lyft drivers, these worker-owners have meaningful control over the conditions of their work — they even helped design the app and the algorithms that coordinate their labor."[61]

Again, such a future might seem fanciful, but Tarnoff asserts that, "It's closer than you think: some of the elements are already emerging in rudimentary form. For example, programmers like Darius Kazemi—the author of a step-by-step guide to starting a small-scale social media site—are building new kinds of online spaces, with a more deliberate and more democratic approach to content moderation. Another example is Mastodon, an open source software project that enables people to run their own social media servers and link them together into a federation."

Tarnoff provides yet another example of "liberating the internet from the constraints of the profit motive" when he points out that, rather than relying on private broadband service providers, "Across the country, hundreds of publicly and cooperatively owned 'community networks' are developing an alternative to the market-first model."[62]

Do these community-operated networks and workers' collectives still seem like extreme, fringe-y options? Does the idea of valuing the welfare of all everyday workers and citizens[n] as much as the profits of a few company owners and shareholders seem too radical?

Here's a thought experiment: imagine that you're a space traveler and you land on a planet in a foreign galaxy. You discover that just eight of its residents have managed to collect more of its wealth than the poorest 3.6 *billion*.[63,o] And then an epidemic ravages the globe, depriving millions of those creatures of their livelihood, yet the richest manage to double their fortunes.[64,p]

If you saw that, would you be able to tell yourself that the citizens of this planet had instituted the most rational, moral, and optimal economic system?

---

[n] And the welfare of all of the Earth's animals and plants.

[o] According to *Forbes*, of the ten richest people on Earth in 2022, seven of them were American tech billionaires. (https://www.forbes.com/billionaires/)

[p] This actually happened here on Earth during the Covid-19 pandemic. (https://www.oxfam.org/en/press-releases/ten-richest-men-double-their-fortunes-pandemic-while-incomes-99-percent-humanity)

Chapter 19
Rising To The Challenge

"In a sense, abolishing Silicon Valley isn't really possible—at least not anytime soon. And yet the point of making the demand is to illustrate that the systems that govern our world are *constructed*—the product of choices by human beings who came before us. Things weren't always like this, and they don't always have to be like this either."[1]

—Wendy Liu

As I've noted throughout this book, I'm heartened by the ways in which the recent techlashes seem to be motivating people—both outside and within Silicon Valley—to recognize that we ought to make substantial changes to the ways in which we create our technologies, and changes to our economic system.

I must admit, though, that my optimism is tempered by several significant factors (aside from the radically shifted political winds post-2024).

As author Michael Lewis put it, in regard to why the world

didn't do a better job of responding to the Covid pandemic, "There is no incentive to prevent things. [...] We've given ourselves over to markets in a way that's pretty extreme. Which is to say, we strongly encourage things that pay and we give correspondingly less attention to things that don't pay. Prevention does not pay."[2]

As I consulted critics like L.M. Sacasas, I heard some skepticism about the notion that we're on the edge of a major improvement in the ways in which we deal with our technologies. "But against what, exactly, is the backlash?" he has written.

> "Is it against the ascendency of technology as the driving principle of modern society? Are we in the midst of a radical reordering of the way individuals relate to the devices, processes, and systems that increasingly order their lives? A closer look will reveal something far more modest: a limited series of reactions to specific cases of carelessness or overtly unethical behavior by tech companies. [...] Although some have said, and not entirely without justification, that this time is different, early indications strongly suggest that the tech companies will successfully navigate the storm and that technology's place in contemporary society will be undisturbed."[3]

Sacasas has also pointed out that even if we change our economic system and make the other changes I have mentioned, there are problems inherent in some of our technologies that might remain. I'm reminded of an interview titled "Why Are American Teenagers So Sad and Anxious?" in which Jonathan Haidt spoke about the "mental health catastrophe unfolding" in our young people, especially adolescent girls. One of the chief reasons he cited is their heavy use of social media.

> "I think the most poisonous, dangerous, damaging process is you post photos of yourself, your body, your face for strangers to rate .

. . and then the pain you feel when people make a critical comment, or when they say nothing at all. So I think what we have here is a platform that is unsafe at any speed. People talk about how to tweak it. 'Well, let's hide the Like counter,' is what Instagram tried. But let me say this very clearly: There is no way, no tweak, no architectural change that will make it okay."[4]

"We want to approach it as a technical problem," Sacasas said, "where if we just create the right filters or the right algorithms, or add the right little tag line to a post warning about its content . . . The question that I would raise is, is it good for us as human beings to be this connected, pervasively and perpetually, to have these flows of information always coming at us indiscriminately, whether or not social media companies, say, do a better job of moderation, whether or not they do a better job of the problem of misinformation?"[5]

Adding a few regulations and coding changes won't necessarily get at the full root of our problems. "We're embedded," he says, "in techno-economic systems that are ordered largely towards the accomplishment of goods and tasks and aims and objectives that are not really related to the well-being of human beings, writ large."[6]

Jaron Lanier expresses this more starkly:

"The business model that makes life worse is based on a particular ideology. This ideology holds that humans as we know ourselves are being replaced by something better that will be brought about by tech companies. Either we'll become part of a giant collective organism run through algorithms, or artificial intelligence will soon be able to do most jobs, including running society, better than people. The overwhelming imperative is to create something like a universally Facebook-connected society or a giant artificial intelligence. These 'new gods' run on data, so as much data as possible must be gathered, and getting in the middle of human interactions is how you gather that data. If the process makes people crazy, that's an acceptable price to pay."[7]

As I have argued all along, what we might need to change first of all is not how we regulate tech, or even how we make it, but how we think of it.

As Sacasas puts it:

"We fail to ask, on a fundamental level, if there are limits appropriate to the human condition, a scale conducive to our flourishing as the sorts of creatures we are. Modern technology tends to encourage users to assume that such limits do not exist; indeed, it is often marketed as a means to transcend such limits. We find it hard to accept limits to what can or ought to be known, to the scale of the communities that will sustain abiding and satisfying relationships, or to the power that we can harness and wield over nature. We rely upon ever more complex networks that, in their totality, elude our understanding, and that increasingly require either human conformity or the elimination of certain human elements altogether. But we have convinced ourselves that prosperity and happiness lie in the direction of limitlessness."[8]

You can find no better illustration of this belief than in J. Storrs Hall's *Where Is My Flying Car?* "Increased productivity," he writes of his vision of the future, "instead of allowing fewer people to produce the same fixed output, can allow everyone to do more and more. This a world of makers instead of takers.[a] This is the

---

[a]   Note the strong echo of Ayn Rand in this phrasing—not to mention other right-wing ideologues, when scorning social welfare programs.

world in which everyone can afford a million-dollar flying car and vacation around the rings of Saturn. This is the world where we build cities on the sea and colonies in space."[9] An explicit running theme of his book is that we really ought to have the world shown in *The Jetsons*. (Hall is clearly brilliant when it comes to engineering issues, yet he models his ideal of the future on a 1960s TV cartoon.)

He also advocates relentlessly for a number of the potentially dangerous technologies we have looked at here—orbital space platforms, geoengineering, nanotechnology, etc.—with little or no consideration of how they might have disastrous effects. He barely takes into account such major real-world forces as political differences, malicious actors, and distorting economic incentives.

He says that those who express concerns are just "virtue-signaling,"[10] "decadent"[11] nemeses of innovation, motivated by "idiotic fear."[12] With unintended irony, he ends his book with a quote from airplane inventor Wilbur Wright, who said that "It is not really necessary to look too far into the future; we see enough already to be certain it will be magnificent. Only let us hurry and open the roads."[13]

Unfortunately, this is the ideology embedded in today's world of tech. It's founded in the notion that creating more advanced technologies always constitutes Progress, and that our only choices are to either get on board, or get out of the way.

But the roads, as we have seen over and over since the time of the Romans, run in more than one direction, and we're not always able to determine which way the traffic will flow.

Unsettled by Sacasas's argument about the limitations of various "fixes" for our errant technologies, and feeling a bit of despair about what we can do, I circled back for a second interview.

I soon learned that the tech critic is not at all nihilistic. He told me that he applauds those who are striving to make changes in the realms of policy, regulation, and law, and who work toward creating alternative economic systems, and he believes such efforts are important. But he said that the question he keeps coming back to, given the reality that the capitalist system is not liable to be dismantled any time soon, is, "What do I do tomorrow?"[14] He said he strives to find "better ways of living my life, ordering my family, and creating stable communities around me," and this involves paying close attention to how he uses technologies.

I pointed out that many big corporations would be only too happy for citizens to take on that responsibility, and that some critics even argue that putting our focus on what we can do as individuals is not only largely ineffective, but may divert us from taking collective action. I mentioned how, for example, the worldwide plastics industry, which produces 300 million tons of that long-lasting material every year,[b] is glad to promote home recycling as a solution, since the onus falls on people in their homes, rather than on companies for creating the problem in the first place. We think we're helping to solve the problem, although this response is far less effective than the plastics and packaging industries would like us to believe.[c]

Sacasas acknowledged the need for collective action, but countered that he "just can't quite abide with the dismissal of the importance of individuals thinking carefully and wisely about the choices they make. I think it matters what person I end up being as a human being, in some ultimate kind of sense, but also for the sake of my ethical and moral responsibilities to my friends, my

---

[b] The National Resources Defense Council points out that this is "nearly equivalent to the weight of the entire human population"—and half of it is single-use. (https://www.nrdc.org/stories/single-use-plastics-101)

[c] According to the NRDC, "a whopping 91 percent of all plastic isn't recycled at all, but [...] ends up in landfills or in the environment."

family and my community. And so I can't just discount that and say it doesn't matter what the individual does, it doesn't matter if you try to consume less, or try to order your life around something other than production. I think it does matter."

Ultimately, we agreed that the conflict between individual and collective action represents a false dichotomy. As Sacasas put it, "It's not an either/or."

In regard to what he can achieve as a tech blogger and critic, he strives to deepen the awareness of his readers. "I have a very limited circle of influence," he told me, "but I think if I'm asked what is the most valuable change that I can effect in somebody, it is to think again about how attention channeled through the body to the world can radically transform the way we live and act in it, can renew us for action, for social and political action, can give us a vision of the good we want to pursue instead of the evil that we always for good reason want to decry. And so, it's rejuvenating our imagination by learning again to see the world and to love it and care for it."

"I know this sounds so sentimental," he added, but I disagree; I think it sounds human-centered and humane, and I'd much rather heed this kind of voice than the recklessly cheery one of the extreme techno-optimist, or the cold, ugly one of the neo-Randian.

When Sacasas talked about "learning again to see the world," I couldn't help thinking about those 17th century Native Americans who went to Europe and were astounded that a society could let people become so impoverished that they would have to sleep in the streets, and not feel any obligation to take care of them. We still have desperately poor people living in the streets of most "First World"[d] cities today. Like those shocked Native Americans, we should never accept such things as normal, as "the way things

---

[d] The term implies that industrialized capitalist countries with the most advanced technologies are also the most civilized.

are." Perhaps our societies that can develop autonomous killing robots and send a few billionaires into space, yet can't provide food and shelter for all our citizens, are not as advanced as we like to think.

~

So how *can* we ensure that our technologies benefit everyone, and our environment as well?

Again, at the root, we need to re-think how we think of them. If we read Mary Shelley's *Frankenstein* as a story about a technology that somehow just ended up going wrong, we're missing the essential point, which is that only *people* make technologies, and that there are often things that are wrong with how and why we make them.

Ultimately, we can't just play defense against tech tycoons and their companies; we can't just be reactive and struggle to protect ourselves from creations that have already been loosed upon us. If we want real change, it has to come at the other end of the process, when people are making the initial decisions to invent things, and drawing up their first designs.

Our focus should not just be on coming up with answers to problems—it has to involve asking the right questions to begin with. Not *Can we quickly scale up this startup so we can sell it and make a killing?* but *Who gets to decide if and how this technology gets made? And Who is this technology really designed to benefit? And Will this technology genuinely contribute to enriching the lives of human beings, and protecting our environment?*

Our challenge is not just to avoid negative unintended consequences. It's to choose what *positive* goals we ultimately have, and to bolster the conscientious and intentional development of technology that will lead us in their direction. We need to be guided by a vision of the collective moral good. As Tim O'Reilly put it, "We can't just expect somehow that the market will magically come up with solutions. We have to rise to the challenges through political leadership, through intellectual leadership, even through religious leadership. We have to have a moral revolution in ourselves, where we come to believe different things about what should be and what will be because we make it so."[15]

This need for a different vision was something Paulina Borsook came to realize after she attended a tech conference in Silicon Valley in 1997. "Missing from the debate," she wrote, "was a sense of what value there might be in giving something up—an illusion of perfect personal freedom, no chains, no claims, no demands, no strictures—in return for gaining something larger, such as connection, commitment, a sense of reliability on the artifice of human society, intimacy and emotional interdependency, and the benefits of generalized, free-floating social contract."[16]

In short, I think we need to start with a fundamental conviction that technology should serve people, and not the other way around—and that we should organize our society around keeping all of us fed, housed, healthy, safe, and happy, not on minting more unicorns and billionaires.

But still: what can you and I do, as lone individuals in this world, faced with the bulldozer might of global corporations?

As individual users of tech, we can decide to become more mindful about our relationship to it. We can reduce our time interacting with those technologies that don't encourage our best behaviors, and don't support our best interests as human beings. In particular, we can resist Big Tech's relentless attempts to monopolize our attention.[17] We can reject the habit of reaching for our phones every time we have a few seconds of downtime. We can just daydream, or allow ourselves to be bored. In 2022, I felt inspired by a group of Brooklyn high schoolers who started a Luddite Club devoted to freeing themselves from their smartphones and social media. They got together and read books, or made art, or just enjoyed each other's IRL company in a local park.[18]

If you have contact with young people, you can do a more balanced job of imparting the history of technology, praising the positive developments, but also acknowledging the downsides. You can help tomorrow's creators and users of tech develop strong ethical principles. Above all, perhaps, you can help them understand that technologies are not faceless forces that are going to inevitably happen *to them*—but rather, that every single technology can only be made by human beings, and involves decisions we should all help make. If you're a parent, you can limit your children's exposure to tech, and make it easier for them to free themselves from addictive technologies by creating communities that will support that goal.

If you're a creator of tech, you can think more carefully about why you're building it, and who it will benefit, and who it might harm. If you're a worker inside a tech company, you can advocate for corporate decisions that aim toward the greater social good, and you can speak out against those that don't. You can organize with your colleagues to give yourselves a stronger voice. If you're still ignored by the powers that be, you can blow the whistle.

If you're a journalist or other writer, you can help expose technologists who are doing things that might have bad consequences, and shine a light on those who are developing saner, healthier products.

If you're a politician, you can write and *actually enact* laws and regulations that will rein in harmful behaviors of tech companies, and incentivize positive ones. If you're a voter, you can support politicians who are willing to do that. And you can vote with your consumer dollars to support the companies that take the healthiest approach.

As a citizen and a worker, you can support labor unions, one of the few forces that are able to effectively counter the staggering power of the big tech companies. The surge in strikes in 2023 showed that employees don't have to just lie down and get steam-rollered by new technologies, but can demand reasonable protections from them, and a fair share of the profits.

What happens next with various technologies will make a huge difference in my life, and in yours. We're all stakeholders here. Global warming, to take just one example, will affect each and every one of us, from that little six-year-old girl in a village in Namibia who might experience droughts to the billionaires living in the Hamptons whose estates might end up under water due to rising sea levels. These issues are too important to be left up to a few tech company boy emperors, and a few venture capitalists, and a few neoliberal politicians.

We should all get a say here, and especially our children, who are going to have to live with the consequences of the technologies we create.

We're *not* powerless.

It's mindboggling to think of how, throughout recorded history, humans have bowed down to and financially supported a tiny number of people who were called kings and queens, simply because they were born into that position. But look at where we are now: we managed to replace many of the monarchies of the world with democratic governments. That was a huge shift, and it goes to show that change *is* possible.

I think of a quote from Graeber and Wengrow, in which they point out that:

> "Any account which appears to show human beings collectively shaping their own destiny, or even expressing freedom for its own

sake, will likely be written off as illusory, awaiting 'real' scientific explanation; or if none is forthcoming, [...] as outside the scope of social theory entirely. This is one reason why most 'big histories' place such a strong focus on technology. Dividing up the human past according to the primary material from which tools and weapons were made (Stone Age, Bronze Age, Iron Age) or else describing it as a series of revolutionary breakthroughs (Agricultural Revolution, Urban Revolution, Industrial Revolution), they then assume the technologies themselves largely determine the shape that human societies will take for centuries to come."[19]

I cherish the findings of those two social scientists and of James C. Scott, who showed that our history is *not* just a repeating saga of humans forced by technological developments into inequitable social and political arrangements, but a tale of many peoples around the world who were able to make fluid, active decisions about how they wanted to live—who actually *lived out* alternatives to the dominant economic, social—and technological—orders of their times.

May we have the courage to do the same.

Acknowledgments

I'm deeply grateful for various sources of support and encouragement while writing this book. These included Pratt Institute (for sabbatical and research support), Chuck Susswein and Wynne Sax-Cedar, Fred Riedel, Roxanne Aubrey, Vicky Bijur, Jonathan Green, Irina Reyn, and Erika Goldman.

My heartfelt thanks to Andrew Boyd, Jessica DuLong, Wayne Grady, Wendy Liu, and Tom Zoellner, excellent writers.

Special thanks to Virginia Vitzthum (stern but helpful taskmaster); my wise and lovely wife Hilary; and the reporters and editors of the *New York Times* and the many other vital sources of fact-checked journalism cited in this book, as well as all those within and outside Silicon Valley who provided useful information and quotes.

A Most Special Thanks to the very smart, insightful people who kindly agreed to be interviewed for the book: Katy Cook, Daniel Deudney, Gary Klein, Noah Kulwin, Steven Johnson, Douglas Rushkoff, and L.M. Sacasas.

About the Author

Gabriel Cohen is a tenured Adjunct Associate Professor in the B.F.A. Writing Program at Pratt Institute in Brooklyn, New York, where he teaches courses in journalism, critical thinking and writing, and fiction. He has also taught writing at New York University, the Center for Fiction, and Long Island University. He is the author of the nonfiction book *Storms Can't Hurt the Sky: A Buddhist Path Through Divorce* and the novels *Red Hook* (which was nominated for an Edgar Award), *The Graving Dock, Neptune Avenue, The Ninth Step,* and *Boombox.* His new novel *Where You'll Find Me* will be published in August of 2026. He has written journalism and essays for the *New York Times, Poets & Writers,* and many other publications, and worked as a staff writer for the *New Haven Advocate* weekly newspaper. He lives in Brooklyn with his wife.

Please visit GabrielCohenBooks.com.

Bibliography

Alter, Adam, *Irresistible: The Rise of Addictive Technology and the Business of Keeping Us Hooked* (New York: Penguin Press, 2017)

Ball, James, *The Tangled Web We Weave: Inside the Shadow System That Shapes the Internet* (Brooklyn: Melville House, 2020)

Borsook, Paulina, *Cyberselfish: A Critical Romp through the Terribly Libertarian Culture of High Tech* (New York: Public Affairs, 2000)

Bostrom, Nick, and Cirkovic, Milan M., eds., *Global Catastrophic Risks* (Oxford: Oxford University Press, 2008)

Boyd, Andrew, *I Want a Better Catastrophe: Navigating the Climate Crisis with Grief, Hope, and Gallows Humor* (Canada: New Society Publishers, 2023)

Burke, James, *Connections: From Ptolemy's Astrolabe to the Discovery of Electricity; How Inventions Are Linked—and How They Cause Change Throughout History* (New York: Simon and Schuster, 2007)

Cook, Katy, *The Psychology of Silicon Valley: Ethical Threats and Emotional Unintelligence in the Tech Industry* (Switzerland: Palgrave Macmillan, 2020)

Carr, Nicholas, *The Shallows: How the Internet Is Changing the Way We Think, Read, and Remember* (Atlantic Books, 2020)

Crawford, Kate, *Atlas of AI* (New Haven: Yale University Press Books, 2021)

Daub, Adrian, *What Tech Calls Thinking* (New York: FSG Originals x Logic, 2020)

Deudney, Daniel, *Dark Skies: Space Expansionism, Planetary Geopolitics, & the Ends of Humanity* (New York: Oxford University Press, 2020)

Diamond, Jared, *Guns, Germs, and Steel* (New York: W.W. Norton, paperback version, 1999)

Doctorow, Cory, *How to Destroy Surveillance Capitalism* (New York: Medium Editions, 2020)

Duggan, Lisa, *Mean Girl: Ayn Rand and the Culture of Greed* (Oakland: University of California Press, 2019)

Fisher, Adam, *Valley of Genius: The Uncensored History of Silicon Valley (As Told by the Hackers, Founders, and Freaks Who Made It Boom)* (New York: Twelve Books, 2018)

Franklin, Ursula, *The Real World of Technology* (House of Anansi Press, 2004; first published 1990)

Bibliography

Gavet, Maëlle, *Trampled by Unicorns: Big Tech's Empathy Problem and How to Fix It* (Hoboken: Wiley, 2021)

Ghosh, Dipayan, *Terms of Disservice: How Silicon Valley is Destructive by Design* (Washington D.C.: Brookings Institution Press)

Goldberg, Natalie, *The Great Spring: Writing, Zen, and This Zigzag Life* (Boulder: Shambhala Publications, 2016)

Grady, Wayne, *Technology* (Toronto: Groundwood Books, 2010)

Graeber, David and Wengrow, David, *The Dawn of Everything: A New History of Humanity* (New York: FSG, 2021)

Haidt, Jonathan, *The Anxious Generation: How the Great Rewiring of Childhood is Causing an Epidemic of Mental Illness* (New York: Penguin Press, 2024)

Hall, J. Storrs, *Where Is My Flying Car?* (San Francisco: Stripe Press, 20210)

Harris, Malcolm, *Palo Alto: A History of California, Capitalism, and the World* (New York: Little, Brown, 2023)

Hayes, Chris, *The Siren's Call: How Attention Became the World's Most Endangered Resource* (New York: Penguin Press, 2025)

Janis, Irving, *Victims of Groupthink: A Psychological Study of Foreign-Policy Decisions and Fiascoes* (Boston: Houghton-Mifflin, 1972)

Johnson, Steven, *Farsighted: How We Make the Decisions That Matter the Most* (Great Britain: John Murray Publishers, 2018)

Johnson, Steven, *Wonderland: How Play Made the Modern World* (New York: Riverhead Books, 2016)

Kahneman, Daniel, *Thinking Fast and Slow* (New York: Farrar Straus and Giroux, 2011)

Kelly, Kevin, *The Inevitable: Understanding the 12 Technological Forces That Will Shape Our Future* (New York, Penguin Books, 2016)

Kelly, Kevin, *What Technology Wants* (New York: Penguin Books, 2011)

Lanier, Jaron, *You Are Not a Gadget* (New York: Knopf, 2010)

Liu, Wendy, *Abolish Silicon Valley: How to Liberate Technology from Capitalism* (London: Repeater Books, 2020)

Martinez, Antonio Garcia, *Chaos Monkeys: Obscene Fortune and Random Failure in Silicon Valley* (New York: HarperCollins, 2016, 2018)

McLuhan, Marshall, *The Gutenberg Galaxy: The Invention of Typographic Man* (University of Toronto Press, 1962)

McNamee, Roger, *Zucked: Waking Up to the Facebook Catastrophe* (New York: Penguin Press, 2019)

McRaney, David, *How Minds Change: The Surprising Science of Belief, Opinion, and Persuasion* (New York: Portfolio/Penguin, 2022)

Mumford, Lewis, *Technics and Civilization* (Chicago: The University of Chicago Press, 1934)

Bibliography

Postman, Neil, *Technopoly: The Surrender of Culture to Technology* (New York: Vintage, 1992)

Pursell, Carroll, *White Heat: People and Technology* (Berkeley: University of California Press, 1994)

Ratey, John G., and Manning, Richard, *Go Wild: Eat Fat, Run Free, Be Social, and Follow Evolution's Other Rules for Total Health and Well-Being* (New York: Little, Brown Spark, 2014)

Reich, Rob and Sahami, Mehran, and Weinstein, Jeremy, *System Error: Where Big Tech Went Wrong and How We Can Reboot* (New York: HarperCollins, 2021)

Robinson, Kim Stanley, *The Ministry for the Future* (New York: Orbit Books, 2021)

Rushkoff, Douglas, *Team Human* (New York: Norton, 2019)

Rushkoff, Douglas, *Throwing Rocks at the Google Bus: How Growth Became the Enemy of Prosperity* (New York: Portfolio/Penguin, 2016)

Scott, James C., *Against the Grain: A Deep History of the Earliest States* (New Haven: Yale University Press, 2017)

Taplin, Jonathan, *Move Fast and Break Things: How Facebook, Google, and Amazon Cornered Culture and Undermined Democracy* (Back Bay Books: 2017)

Tarnoff, Ben, and Weigel, Moira, *Voices from the Valley* (New York: FSG Originals x Logic, 2020)

Taylor, Astra, *The People's Platform: Taking Back Power and Culture in the Digital Age* (New York: Picador, 2014)

Tetlock, Philip and Gardner, Dan, *Superforecasting: The Art and Science of Prediction* (New York: Broadway Books, 2015)

Wallach, Wendell, *A Dangerous Master: How to Keep Technology from Slipping Beyond Our Control* (New York: Basic Books, 2015)

Weinersmith, Kelly and Weinersmith, Zach, *A City on Mars: Can We Settle Space, Should We Settle Space, and Have We Really Thought This Through?* (Penguin Press, 2023)

Winner, Langdon, *Autonomous Technology: Technics-out-of-Control as a Theme in Political Thought,* (Cambridge: MIT Press, 1978)

Winner, Langdon, *The Whale and the Reactor: A Search for Limits in an Age of High Technology* (Chicago: The University of Chicago Press, 1989)

Wu, Tim, *The Master Switch: the Rise and Fall of Information Empires* (New York: Vintage Books, 2011)

Zuboff, Shoshana, *The Age of Surveillance Capitalism: The Fight for A Human Future at the New Frontier of Power* (New York: Public Affairs, 2019)

Notes

"Fix"

1. These definitions of "fix" come from *Merriam-Webster's Collegiate Dictionary, Eleventh Edition* (Merriam-Webster Inc., 2020) p. 474

Introduction: On Sharks & Killer Coconuts

1. "2023 Expert Survey on Progress in AI," Aug. 17, 2023 from AI Impacts, https://wiki.aiimpacts.org/ai_timelines/predictions_of_human-level_ai_time lines/ai_timeline_surveys/2023_expert_survey_on_progress_in_ai
2. Yuval Harari, Tristan Harris, and Aza Raskin, "You Can Have the Blue Pill or the Red Pill, and We're Out of Blue Pills," Mar. 24, 2023 in the *New York Times*, https://www.nytimes.com/2023/03/24/opinion/yuval-harari-ai-chatg pt.html
3. Roger McNamee, *Zucked: Waking Up to the Facebook Catastrophe* (New York: Penguin Press, 2019) pp. 16-17
4. From an interview by Scott Pelley on *Sixty Minutes* on Oct. 3, 2021, https:// www.rev.com/blog/transcripts/facebook-whistleblower-frances-haugen-60-minutes-interview-transcript
5. The statistic is for fatalities from unprovoked attacks by sharks. See "Yearly Worldwide Shark Attack Summary" from the International Shark Attack File of the Florida Museum of Natural History, 2023, https://www.floridamu seum.ufl.edu/shark-attacks/yearly-worldwide-summary/ ; also "Shark Death Statistics" June 17, 2024 on WorldMetrics.org, https://worldmetrics.org/ shark-death-statistics/
6. *Marine Policy*, Volume 40, July 2013. pp. 194-204
7. Gabriel Cohen, "Shark Fight!" *New York Post Page Six* magazine, Aug. 31, 2008,https://www.gabrielcohenbooks.com/attachments/G--Cohen-Shark-Fight.pdf
8. Cecil Adams, "Are 150 People Killed Each Year by Falling Coconuts?" July 19, 2002 in *The Straight Dope*, https://www.straightdope.com/21343538/ are-150-people-killed-each-year-by-falling-coconuts
9. From a transcript on *Time.com*, Aug. 23, 2017, https://time.com/4912055/ donald-trump-phoenix-arizona-transcript/
10. Marshall McLuhan, *The Gutenberg Galaxy: The Invention of Typographic Man.* (University of Toronto Press, 1962)

11. Neil Postman, *The Disappearance of Childhood*, (New York: Delacorte Press, 1982) p. 23

12. James Burke, *Connections: From Ptolemy's Astrolabe to the Discovery of Electricity; How Inventions Are Linked—and How They Cause Change Throughout History* (New York: Simon and Schuster, 2007) pp. 204-207

13. Burke, ibid., p. 207

14. Burke, ibid., pp. 278-283

15. Steven Johnson, *Farsighted: How We Make the Decisions That Matter the Most* (Great Britain: John Murray Publishers, 2018) p. 86

16. Roger McNamee, ibid., pp. 4-7

17. Interview with the author, March 13, 2022

18. Allison C. Meier, "An Affordable Radio Brought Nazi Propaganda Home," Aug. 30, 2018 in *JSTOR Daily*, https://daily.jstor.org/an-affordable-radio-brought-nazi-propaganda-home/

19. http://durenberger.com/wp-content/uploads/2018/08/GOEBBELS.pdf

20. I found this analogy in Daniel Deudney, *Dark Skies: Space Expansionism, Planetary Geopolitics, & the Ends of Humanity* (New York: Oxford University Press, 2020) p. 274

21. Charlotte Higgins, "What's Behind Mark Zuckerberg's Man-Crush on Emperor Augustus," Sept. 12, 2018, on *TheGuardian.com*, https://www.theguardian.com/commentisfree/2018/sep/12/what-attracts-mark-zuckerberg-roman-hardman-augustus

22. Christopher Klein, "How Ancient Rome Thrived During Pax Romana," June 8, 2023 on *History.com*, https://www.history.com/news/pax-romana-roman-empire-peace-augustus

23. Erin Griffith, "Facebook's New Mission: Video Will Bring Us Together," on Wired.com, Nov. 3, 2017, https://www.wired.com/story/facebooks-new-mission-video-will-bring-us-together/

1. "This Changed Everything": The Advent of Internet Advertising

1. Jennifer Egan, *The Candy House* (New York: Scribner, 2022) p. 125

2. Maria Popova, "Steve Jobs on Why Computers Are Like a Bicycle for the Mind (1990)," Dec. 21, 2011 on *TheMarginalian.org*, https://www.themarginalian.org/2011/12/21/steve-jobs-bicycle-for-the-mind-1990/

3. Paulina Borsook, *Cyberselfish: A Critical Romp through the Terribly Libertarian Culture of High Tech* (New York: Public Affairs, 2000) p. 223

4. Interview with the author on April 10, 2023

5. Steven Johnson, "Beyond the Bitcoin Bubble," in the *New York Times Magazine*, Jan. 16, 2018, https://www.nytimes.com/2018/01/16/magazine/beyond-the-bitcoin-bubble.html

6. McNamee, *Zucked.*, p. 37

7. Ben Tarnoff, "How the Internet Was Invented," in *The Guardian*, 15 July 15, 2016, https://www.theguardian.com/technology/2016/jul/15/how-the-internet-was-invented-1976-arpa-kahn-cerf

8. Paulina Borsook, ibid., pp. 229-230

9. Tim Wu, *The Master Switch: the Rise and Fall of Information Empires* (New York: Vintage Books, 2011) p. 6

10. Steven Johnson, "Beyond the Bitcoin Bubble," ibid.

11. Sergey Brin and Lawrence Page, "The Anatomy of a Large-Scale Hypertextual Web Search Engine," http://infolab.stanford.edu/~backrub/google.html

12. Shoshana Zuboff, *The Age of Surveillance Capitalism: The Fight for A Human Future at the New Frontier of* Power (New York: Public Affairs, 2019) p. 84

13. "The Internet: Looking Back on How We Got Connected to the World," on the website of the Federal Communications Commission, https://transition.fcc.gov/omd/history/internet/documents/newsletter.pdf

14. Justin Driskill, "Online Advertising Answers: September 2020," on *TheOnlineAdvertisingGuide.com*, https://theonlineadvertisingguide.com/online-advertising-answers/online-advertising-answers-september-2020/

15. "First Commercial Spam" on *WebArchive.com*, https://web.archive.org/web/20060326032706/http://www.mailmsg.com/SPAM_history_001.htm

16. Adam Fisher, *Valley of Genius: The Uncensored History of Silicon Valley (As Told by the Hackers, Founders, and Freaks Who Made It Boom)* (New York: Twelve Books, 2018) p. 347

17. Quoted in Noah Kulwin, "The Organic Side, to Me, Is Scarier Than the Ad Side," April 2018 in *New York* magazine, https://nymag.com/intelligencer/2018/04/antonio-garcia-martinez-former-facebook-employee-interview.html

18. Quoted in James Ball, *The Tangled Web We Weave: Inside the Shadow System That Shapes the Internet* (Brooklyn: Melville House, 2020), pp. 123-124

19. Geoffrey A. Fowler, "There's No Escape from Facebook Even if You Don't Use It," Aug. 29, 2021 in the *Washington Post*, https://www.washingtonpost.com/technology/2021/08/29/facebook-privacy-monopoly/

20. Salvador Rodriguez, "Facebook Closes Above \$1 Trillion Market Cap for the First Time," June 28, 2021 on *CNBC.com*, https://www.cnbc.com/2021/06/28/facebook-hits-trillion-dollar-market-cap-for-first-time.html

21. Zuboff, ibid., p. 93

22. Astra Taylor, *The People's Platform: Taking Back Power and Culture in the Digital Age* (New York: Picador, 2014) p. 202

23. Taylor, ibid.

24. https://www.britannica.com/topic/Sears-Roebuck-and-Company

25. "Twitter's Growth Pales in Comparison to Facebook's," *Statista.com*, https://cdn.statcdn.com/Infographic/images/normal/3200.jpeg

26. "Number of Monthly Active Facebook Users Worldwide as of 4th Quarter

2023," *Statista.com*, https://www.statista.com/statistics/264810/number-of-monthly-active-facebook-users-worldwide/

27. Dominic Rushe, "WhatsApp: Facebook Acquires Messaging Service in $19bn Deal," Feb. 20, 2014 in *The Guardian*, https://www.theguardian.com/technology/2014/feb/19/facebook-buys-whatsapp-16bn-deal

2. Clicks And Consequences: Psychological And Social Effects

1. Bo Burnham, *Inside (The Songs)* CD, Bo Burnham/IMPERIAL, 2021

2. See the video on Erica Pandey, "Sean Parker: Facebook Was Designed to Exploit Human 'Vulnerability'," Nov. 9, 2017 on *Axios.com*, https://www.axios.com/2017/12/15/sean-parker-facebook-was-designed-to-exploit-human-vulnerability-1513306782

3. Chris Hayes, *The Siren's Call: How Attention Became the World's Most Endangered Resource* (New York: Penguin Press, 2025) p. 153

4. Andrew K. Przybylski and Netta Weinstein, "Can You Connect with Me Now? How the Presence of Mobile Communication Technology Influences Face-to-Face Conversation Quality," July 19, 2012 in *Journal of Social and Personal Relationships*, 30(3), pp. 237-246, https://doi.org/10.1177/0265407512453827

5. Azadeh Leland, Kamran Tavakol, Joel Scholten, Debra Mathis, David Maron, and Simin Bakhshi, "The Role of Dual Tasking in the Assessment of Gait, Cognition and Community Reintegration of Veterans with Mild Traumatic Brain Injury," Dec. 2017 in *MateriaSocioMedica*, https://www.ncbi.nlm.nih.gov/pmc/articles/PMC5723199/; and Kevin P. Madore and Anthony D. Wagner, "Multicosts of Multitasking," Mar-Apr 2019 in *Cerebrum*, https://www.ncbi.nlm.nih.gov/pmc/articles/PMC7075496/

6. Natalie Goldberg, *The Great Spring: Writing, Zen, and This Zigzag Life* (Boulder: Shambhala Publications, 2016) pp. 195-196

7. Hanna Rosin, "The Smartphone Kids Are Not All Right," Mar. 21, 2024 in *TheAtlantic.com*, https://www.theatlantic.com/podcasts/archive/2024/03/smartphone-anxious-generation-mental-health/677817/

8. Jaron Lanier, *You Are Not a Gadget* (New York: Knopf, 2010) p. 70

9. Jonathan Haidt, *The Anxious Generation: How the Great Rewiring of Childhood is Causing an Epidemic of Mental Illness* (New York: Penguin Press, 2024),

10. Haidt, ibid. p. 117

11. Haidt, ibid., p. 34

12. McNamee, *Zucked.*, p. 88

13. Various reporters, "The Facebook Files: A Wall Street Journal Investigation," Oct. 1, 2021 in *the Wall Street* Journal, https://www.wsj.com/articles/the-facebook-files-11631713039?mod=bigtop-breadcrumb

Notes

14. Cristiano Lima-Strong and Naomi Nix, "Zuckerberg 'Ignored' Executives on Kids' Safety, Unredacted Lawsuit Alleges," Nov. 8, 2023 on *WashingtonPost.com*, https://www.washingtonpost.com/technology/2023/11/08/zuckerberg-meta-lawsuit-kids-safety/

15. Katy Cook, *The Psychology of Silicon Valley: Ethical Threats and Emotional Unintelligence in the Tech Industry* (Switzerland: Palgrave Macmillan, 2020), p. 202

16. Cook, ibid., p. 215

17. Cook, ibid., p. 211

18. Geoffrey A. Fowler, "There's No Escape from Facebook Even if You Don't Use It," Aug. 29, 2021 in the *Washington Post*, https://www.washingtonpost.com/technology/2021/08/29/facebook-privacy-monopoly/

19. Zuboff, *The Age of Surveillance Capitalism*, pp. 234-236

20. Dipayan Ghosh, *Terms of Disservice: How Silicon Valley is Destructive by Design*, (Washington D.C.: Brookings Institution Press) p. 178

21. Vauhini Vara, "OpenAI Seems to Be Making a Very Familiar, Very Cynical Choice," June 11, 2025 on *NYTimes.com*, https://www.nytimes.com/2025/06/11/opinion/open-ai-big-tech-advertising.html

22. Zuboff, ibid., p. 288

23. Lindsay Tohana, "Advantages of Holograms and What It Means for the Future of Marketing," Sept. 23, 2020 on *FutureOfMarketingInstitute.com*, https://futureofmarketinginstitute.com/advantages-of-holograms-and-what-it-means-for-the-future-of-marketing/

24. "Total Surveillance is Not What America Signed Up For," by the Editorial Board on *NYTimes.com*, Dec. 21, 2019, https://www.nytimes.com/interactive/2019/12/21/opinion/location-data-privacy-rights.html

3. Cozy Communities And Creepy Clusters: Political Effects

1. Lewis Mumford, *Technics and Civilization* (Chicago: The University of Chicago Press, 1934) p. 241

2. "Understanding QAnon's Connection to American Politics, Religion, and Media Consumption," May 27, 2021, on *PRRI.org* (from the Public Religion Research Institute), https://www.prri.org/research/qanon-conspiracy-american-politics-report/ ; see also Antonia Noori Farzan, "NOT TRUE: Oprah Winfrey Debunks Bizarre QAnon Conspiracy Theory Spreading Across the Internet," March 18, 2020 on *WashingtonPost.com*, https://www.washingtonpost.com/nation/2020/03/18/oprah-winfrey-qanon-conspiracy/

3. "The Influenza Pandemic of 1918,"https://virus.stanford.edu/uda/

4. "Covid Data Tracker," as of March 15, 2025, on *Covid.CDC.gov*, https://covid.cdc.gov/covid-data-tracker/#trends_totaldeaths_select_00

5. See, for example, Jacob Wallace, Paul Goldsmith-Pinkham, and Jason L. Schwartz, "Excess Death Rates for Republican and Democratic Registered Voters in Florida and Ohio During the Covid-19 Pandemic," in *JAMA Intern Med.* 2023;183(9):916-923. doi:10.1001/jamainternmed.2023.1154

6. Linley Sanders, "The Difference Between What Republicans and Democrats Believe to Be True About Covid-19," May 26, 2020 on *YouGov*, https://today.yougov.com/politics/articles/29917-republicans-democrats-misinformation

7. Dan Evon, "Did Stanley Kubrick Fake the Moon Landings?" Dec. 11, 2015 on *Snopes.com*, https://www.snopes.com/fact-check/false-stanley-kubrick-faked-moon-landings/

8. "Separating Fact from Fiction: The Reptilian Elite," July 20, 2009 on *Time.com*, https://content.time.com/time/specials/packages/article/0,28804,1860871_1860876_1861029,00.htm

9. Daniel Kahneman, *Thinking Fast and Slow* (New York: Farrar Straus and Giroux, 2011)pp. 119-128

10. Wikipedia, "List of Cognitive Biases," https://en.wikipedia.org/wiki/List_of_cognitive_biases

11. Wikipedia, ibid.

12. Adrian Daub, *What Tech Calls Thinking* (New York: FSG Originals x Logic, 2020) p. 90

13. Cory Doctorow, *How to Destroy Surveillance Capitalism* (New York: Medium Editions, 2020) p. 25

14. Facebook researcher Monica Lee quoted in Karen Hao, "How Facebook Got Addicted to Spreading Misinformation" Mar. 11, 2021, in *MIT Technology Review*,https://www.technologyreview.com/2021/03/11/1020600/facebook-responsible-ai-misinformation/

15. Joshua Kaplan, "Armed and Underground: Inside the Turbulent, Secret World of an American Militia," Aug. 17, 2024 on ProPublica, https://www.propublica.org/article/inside-secret-ap3-militia-american-patriots-three-percent

16. Sally Younger, "NASA Analysis Confirms Year of Monthly Temperature Records," June 11, 2024 on *NASA.gov*, https://www.nasa.gov/earth/nasa-analysis-confirms-a-year-of-monthly-temperature-records/

17. Sandee LaMotte, "Social Media's Message about the Sun and Sunscreen: 'It's Frightening,'" June 21, 2024 on *CNN.com*, https://www.cnn.com/2024/06/21/health/tiktok-falsehoods-sun-protection-wellness/index.html

18. Nir Grinberg, Kenneth Joseph, Lisa Friedland, Briony Swire-Thompson, and David Lazer, "Fake News on Twitter During the 2016 U.S. Presidential Election," Jan. 25, 2019, in *Science.* https://www.science.org/doi/10.1126/science.aau2706

19. "The Disinformation Dozen: Why Platforms Must Act on Twelve Leading Online Anti-Vaxxers," Mar. 24, 2021 from The Center for Countering Digital Hate, https://counterhate.com/research/the-disinformation-dozen/

20. Laura Edelson, Minh-Kha Nguyen, Ian Goldstein, Oana Goga, Damon Mc-Coy, and Tobias Lauinger, "Understanding Engagement with U.S. (Mis)Information News Sources on Facebook," November 2, 2021, in a joint study conducted by New York University and Université Grenoble Alpes, in ACM Internet Measurement Conference (IMC '21), November 2–4, 2021, Virtual Event, https://dl.acm.org/doi/pdf/10.1145/3487552.3487859

21. Zeynep Tufekci, "YouTube, the Great Radicalizer," Mar. 10, 2018 on *NYTimes.com*,https://www.nytimes.com/2018/03/10/opinion/sunday/youtube-politics-radical.html

22. Tufekci, ibid; see also Jack Nicas, "How YouTube Drives People to the Internet's Darkest Corners," Feb. 17, 2018 in the *Wall Street Journal*, https://www.wsj.com/articles/how-youtube-drives-viewers-to-the-internets-darkest-corners-1518020478

23. Quoted in Noah Kulwin, "The Internet Apologizes for Violating and Hijacking Our Attention," April 16, 2018 in *New York*, https://nymag.com/intelligencer/2018/04/an-apology-for-the-internet-from-the-people-who-built-it.html

24. See, for example, Laura Paisley, "Political Polarization at Its Worst Since the Civil War," Nov. 8, 2016 on *Today.usc.edu*, https://today.usc.edu/political-polarization-at-its-worst-since-the-civil-war-2/

25. McNamee, *Zucked.*, p. 93

26. Devi Sridhar, "The Appointment of Robert F. Kennedy Has Horrified Public Health Experts. Here Are His Three Most Dangerous Ideas," Nov. 17, 2024 in *The Guardian*, https://www.theguardian.com/commentisfree/2024/nov/17/robert-f-kennedy-public-health-usa-donald-trump-most-dangerous-ideas

27. Hayes, *The Siren's Call*, p. 243

28. David McRaney, *How Minds Change: The Surprising Science of Belief, Opinion, and Persuasion* (New York: Portfolio/Penguin, 2022) p. 291

29. Co-author Zeve Sanderson, quoted in "Evaluating the Truthfulness of Fake News Through Online Searches Increases the Chances of Believing Misinformation," Dec. 20, 2023 from NYU's Center for Social Media and Politics, https://csmapnyu.org/impact/news/evaluating-the-truthfulness-of-fake-news-through-online-searches-increases-the-chances-of-believing-misinformation

30. To get a good sense of how all this works for flat Earthers, watch *Behind the Curve*, a 2018 documentary film directed by Daniel J. Clark.

31. Colby Itkowitz, Emily Guskin, and Scott Clement, "Trump Trusted More Than Biden on Democracy Among Key Swing-State Voters," June 26, 2024 on *WashingtonPost.com*, https://www.washingtonpost.com/politics/2024/06/26/biden-trump-swing-state-poll-democracy/

32. Ben Kamisar, "Poll: Biden and Trump Supporters Sharply Divided by the Media They Consume," Apr. 29, 2024 on *NBCNews.com*, https://www.nbcnews.com/politics/2024-election/poll-biden-trump-supporters-sharply-divided-media-consume-rcna149497

33. Asmongold (real name Zach Hoyt), "Legacy Media Realizes It's Dead," Nov. 12, 2024, https://www.youtube.com/live/Dse_hniElxE

34. "Working at the Washington Post," updated Mar. 14, 2024 on *Zippia.com*, https://www.zippia.com/the-washington-post-careers-62401/

35. Elahe Izadi, "After Non-Endorsement, 250,000 Subscribers Cancel the Washington Post," Oct. 29, 2024 on *WashingtonPost.com*, https://www.washingtonpost.com/style/media/2024/10/29/washington-post-cancellations-number/

36. Brett Cooper, "Why I Voted for Donald Trump," on *The Comments Section with Brett Cooper*, Nov. 7, 2024, YouTube, https://www.youtube.com/watch?v=ybgV3jSNKEE

37. Galen Stocking, Luxuan Wang, Michael Lipka, Katerina Eva Matsa, Regina Widjaya, Emily Tomasik, and Jacob Liedke, "America's News Influencers," Nov 18, 2024, from Pew Research Center, https://www.pewresearch.org/journalism/2024/11/18/americas-news-influencers

38. See, for example, Robin Brooks, Peter R. Orszag, and William E. Murdock III, "Covid-19 Inflation Was a Supply Shock," Aug. 15, 2024 from the Brookings Institution,https://www.brookings.edu/articles/covid-19-inflation-was-a-supply-shock/

39. See, for example, Weihua Li and Jamiles Lartey, "Crime Rates and the 2024 Election: What You Need to Know," June 27, 2024 on *TheMarshallProject.org*,https://www.themarshallproject.org/2024/06/27/trump-and-biden-spar-over-crime-rates-ahead-of-their-debate-what-do-we-really-know

40. See, for example, Shane Goldmacher, "How Trump Targeted Undecided Voters Without Breaking the Bank," Dec. 5, 2024 on *NYTimes.com*, https://www.nytimes.com/2024/12/05/us/politics/trump-streaming-ads-strategy.html; and Katherine Fung, "Democrats Hoped the Bros Wouldn't Show. But They Did," Nov. 8, 2024 on *Newsweek.com*, https://www.newsweek.com/donald-trump-young-men-bro-vote-1982213

41. Reuters reported that Trump won 49.9% of the 2024 popular vote, in "U.S. Presidential Election Results," updated Dec. 6, 2024 on *Reuters*.com, https://www.reuters.com/graphics/USA-ELECTION/RESULTS/zjpqnemxwvx/president/

42. As defined in the Wikipedia article on the practice, https://en.wikipedia.org/wiki/Scapegoating

43. Gregory A. Smith, "White Protestants and Catholics Support Trump, But Voters in Other U.S. Religious Groups Prefer Harris," Sept. 9, 2024 on *PewResearch.org*,https://www.pewresearch.org/short-reads/2024/09/09/white-protestants-and-catholics-support-trump-but-voters-in-other-us-religious-groups-prefer-harris/

44. David French, "How a German Thinker Explains MAGA Morality," Jan. 26, 2025 on *NYTimes.com*, https://www.nytimes.com/2025/01/26/opinion/trump-maga-schmitt.html

45. Ghosh, *Terms of Disservice.*, pp. 8-11

46. Ghosh, ibid., p. 11

47. See Julie Carrie Wong, "How Facebook Let Fake Engagement Distort Global Politics: A Whistleblower's Account," Apr. 12, 2021 in the *Guardian*.com, https://www.theguardian.com/technology/2021/apr/12/facebook-fake-engagement-whistleblower-sophie-zhang

48. Craig Silverman, Ryan Mac, and Pranav Dixit, "I Have Blood on My Hands: A Whistleblower Says Facebook Ignored Global Political Manipulation," Sept. 14, 2020 on *BuzzFeedNews.com*, https://www.buzzfeednews.com/article/craigsilverman/facebook-ignore-political-manipulation-whistleblower-memo

49. Cook cites the *V-Dem Annual Democracy Report 2018* from the University of Gothenburg's Varieties of Democracy Institute, https://www.v-dem.net/static/website/files/dr/dr_2018.pdf

50. For more specifics on Facebook's impact on politics, see Alexis C. Madrigal, "What Facebook Did to American Democracy," Oct. 12, 2017 on *TheAtlantic.com*, https://www.theatlantic.com/technology/archive/2017/10/what-facebook-did/542502/

51. Cook, *The Psychology of Silicon Valley*, p. 157

52. Naomi Nix and Sarah Ellison, "Following Elon Musk's Lead, Big Tech Is Surrendering to Disinformation," Sept. 1, 2023 on *WashingtonPost.com*, https://www.washingtonpost.com/technology/2023/08/25/political-conspiracies-facebook-youtube-elon-musk/

53. Kanishka Singh and Sheila Dang, "Musk and X Are Epicenter of US Election Misinformation, Experts Say," Nov. 4, 2024 on *Reuters.com*, https://www.reuters.com/world/us/wrong-claims-by-musk-us-election-got-2-billion-views-x-2024-report-says-2024-11-04/

54. Miles Klee, "How Elon Musk and X Became the Biggest Purveyors of Online Misinformation," Aug. 9, 2024 on *RollingStone.com*, https://www.rollingstone.com/culture/culture-features/elon-musk-twitter-misinformation-timeline-1235076786/

55. Joel Kaplan, "More Speech and Fewer Mistakes," Jan. 7, 2025 on *About.F-B.com*, https://about.fb.com/news/2025/01/meta-more-speech-fewer-mistakes/

56. Kate Knibbs, "Meta Now Lets Users Say Gay and Trans People Have 'Mental Illness'," Jan. 7, 2025 on *Wired.com*, https://www.wired.com/story/meta-immigration-gender-policies-change/

57. Clare Duffy, "Calling Women 'Household Objects' Now Permitted on Facebook after Meta Updated Its Guidelines," Jan. 8, 2025 on CNN.com, https://www.cnn.com/2025/01/07/tech/meta-hateful-conduct-policy-update-fact-check/index.html

58. Nico Grant and Tripp Mickle, "YouTube Loosens Rules Guiding the Moderation of Videos," June 9, 2015 on *NYTimes.com*, https://www.nytimes.com/2025/06/09/technology/youtube-videos-content-moderation.html

59. Grant and Mickle, ibid.

60. Quoted in Noah Kulwin, "You Have a Persuasion Engine Unlike Any Created in History," June 2018 in *New York,* https://nymag.com/intelligencer/2018/04/roger-mcnamee-early-facebook-investor-interview.html

61. Quoted in Kara Swisher, "The Metaverse: Expectations and Reality" Nov. 11, 2021 in the *New York Times,* https://www.nytimes.com/2021/11/11/opinion/sway-kara-swisher-jaron-lanier.html

62. See, for example, Isabelle Qian, Muyi Xiao, Paul Mozur, and Alexander Cardia, "Four Takeaways from a Times Investigation into China's Expanding Surveillance State," June 21, 2022 on *NYTimes.com,* https://www.nytimes.com/2022/06/21/world/asia/china-surveillance-investigation.html

63. Zuboff, *The Age of Surveillance Capitalism,* p. 360

4. What If Different Choices Had Been Made?

1. Statistic from Sandvine, a network monitoring company, https://www.sandvine.com/phenomena

2. Zuboff, ibid., p. 85

3. 47 U.S. Code Section 230—Protection for Private Blocking and Screening of Offensive Material, https://www.govinfo.gov/content/pkg/USCODE-2020-title47/pdf/USCODE-2020-title47-chap5-subchapII-partI-sec230.pdf

4. "Section 230 of the Communications Decency Act," https://www.eff.org/issues/cda230

5. Steven Johnson, "Beyond the Bitcoin Bubble," Jan. 16, 2018 in the *New York Times,*https://www.nytimes.com/2018/01/16/magazine/beyond-the-bitcoin-bubble.html

6. Fisher *Valley of Genius,* pp. 145-147

7. Gabriel Cohen, "You Talkin' to Me?" Jan. 9, 2009 in the *New York Times,* https://www.nytimes.com/2009/01/11/nyregion/thecity/11blog.html

8. S. Dixon, "Facebook: Face Account Removal Q4 2017-Q2 2022," Nov. 11, 2022 on *Statista.com,* https://www.statista.com/statistics/1013474/facebook-fake-account-removal-quarter/

9. See, for example, John Herrman, "You Anon: Reconsidering Pseudonymity and What It Means to 'Be Yourself' Online," July 31, 2021 in the *New York Times,*https://www.nytimes.com/2021/07/31/style/anonymity-pseudonymity-online-identity.html

10. Gilad Edelman, "Everything You've Heard About Section 230 Is Wrong," May 6, 2021 on *Wired.com,* https://www.wired.com/story/section-230-internet-sacred-law-false-idol/

11. Quoted in Noah Kulwin, "The Internet Apologizes for Violating and Hijacking Our Attention," April 16, 2018 in *New York,* https://nymag.com/intelligencer/2018/04/an-apology-for-the-internet-from-the-people-who-built-it.html

12. Edelman, ibid.

13. Edelman, ibid.

14. "Section 230," https://en.wikipedia.org/wiki/Section_230#Immunity_from_content_related_to_illegal_activities

15. Ball, *The Tangled Web We Weave*, p. 104

16. https://www.britannica.com/topic/British-Broadcasting-Corporation

17. Wu, *The Master Switch* p. 74

18. Kulwin, ibid.

19. Adam Briggle, "As Frankenstein Turns 200, Can We Control Our Modern 'Monsters'?" Dec. 29, 2017 on *ScientificAmerican.com*, https://www.scientificamerican.com/article/as-frankenstein-turns-200-can-we-control-our-modern-monsters/

20. Noah Kulwin, "I Fundamentally Believe that My Time at Reddit Made the World a Worse Place," Apr. 16, 2018 in *New York*, https://nymag.com/intelligencer/2018/04/dan-mccomas-reddit-product-svp-and-imzy-founder-interview.html

5. "The Sword And The Spear": The Danger Of Technological Arms Races

1. Carroll Doherty and Jocelyn Kiley, "Americans Have Become Much Less Positive About Tech Companies' Impact on the U.S," July 29, 2019 on *PewResearch.org*,https://www.pewresearch.org/short-reads/2019/07/29/americans-have-become-much-less-positive-about-tech-companies-impact-on-the-u-s/

2. Meghan Brenan, "Views of Big Tech Worsen; Public Wants More Regulation," Feb. 18. 2021 from Gallup, Inc., https://news.gallup.com/poll/329666/views-big-tech-worsen-public-wants-regulation.aspx

3. See, for example, Matthew Rosenberg, Nicholas Confessore, and Carole Cadwalladr, "How Trump Consultants Exploited the Facebook Data of Millions," March 17, 2018 in the *New York Times*, https://www.nytimes.com/2018/03/17/us/politics/cambridge-analytica-trump-campaign.html

4. David Hamilton, "Mark Zuckerberg's Long Apology Tour: A Brief History," Feb. 1, 2024 on *APNews.com*, https://apnews.com/article/mark-zuckerberg-facebook-apology-c7055b654f63a23d09b6a96388dfa2b4

5. Mark Zuckerberg, "Founder's Letter, 2021," Oct. 28, 2021 on *About.fb.com*, https://about.fb.com/news/2021/10/founders-letter/

6. Sheera Frenkel and Kellen Browning "The Metaverse's Dark Side: Here Come Harassment and Assaults," Dec. 30, 2021 on *NYTimes.com*, https://www.nytimes.com/2021/12/30/technology/metaverse-harassment-assaults.html

7. Danielle Keats Citron, "The Continued (In)visibility of Cyber Gender Abuses," Nov. 22, 2023 in the *Yale Law Journal*, Volume 133 202302024, https://

www.yalelawjournal.org/forum/the-continued-invisibility-of-cyber-gender-abuse#_ftnref144

8. "(FB) CEO Mark Zuckerberg on Q2 2021 Results - Earnings Call Transcript" from an Earnings Conference call on July 28, 2021, https://seekingalpha.com/article/4442353-facebook-inc-fb-ceo-mark-zuckerberg-on-q2-2021-results-earnings-call-transcript

9. Yasmin Khorram, "Meta's Reality Check: Inside the $45 Billion Cash Burn at Reality Labs," July 28, 2024 on *Finance.yahoo.com,* https://finance.yahoo.com/news/metas-reality-check-inside-the-45-billion-cash-burn-at-reality-labs-125717347.html

10. John Herrman, "How Mark Zuckerberg Led the Tech Industry Into a Metaverse Wasteland," May 10, 2023 in *New York,* https://nymag.com/intelligencer/2023/05/the-metaverse-was-a-ridiculous-idea-where-did-it-come-from.html

11. Ed Zitron, "RIP Metaverse: An Obituary for the Latest Fad to Join the Tech Graveyard," May 8, 2023 on *BusinessInsider.com,* https://www.businessinsider.com/metaverse-dead-obituary-facebook-mark-zuckerberg-tech-fad-ai-chatgpt-2023-5

12. Kali Hays and Ashley Stewart, "Mark Zuckerberg's Metaverse Obsession Is Driving Some Current and Former Facebook Employees Nuts: 'It's the Only Think Mark Wants to Talk About," Apr. 22, 2022 in *BusinessInsider.com,* https://www.businessinsider.com/mark-zuckerberg-metaverse-obsession-driving-some-employees-nuts-2022-4

13. See, for example, Nitish Pahwa, "Why the Metaverse Has to Look So Stupid," Aug. 19, 2022 on *Slate.com,* https://slate.com/technology/2022/08/mark-zuckerberg-metaverse-horizon-worlds-facebook-looks-crappy-explained.html

14. See, for example, Ben Hartwig, "Top 11 Benefits of Artificial Intelligence in 2024," Jan. 4, 2023 on *Hackr.io,* https://hackr.io/blog/benefits-of-artificial-intelligence

15. "Science AMA Series: Stephen Hawking AMA Answers!" July 27, 2015 on Reddit.com, https://www.reddit.com/r/science/comments/3nyn5i/science_ama_series_stephen_hawking_ama_answers/

16. Nick Bostrom, "Existential Risks: Analyzing Human Extinction Scenarios and Related Hazards," in the *Journal of Evolution and Technology,* Vol. 9, No. 1 (2002)

17. Kathleen Miles, "Artificial Intelligence May Doom the Human Race Within a Century, Oxford Professor Says," Feb. 4, 2015 on *HuffPost.com,* https://www.huffpost.com/entry/artificial-intelligence-oxford_n_5689858

18. Isaac Asimov, introduced in his short story "Runaround," published in the collection *I, Robot* (New York: Doubleday, 1950)

19. Katja Grace, Harlan Stewart, Julia Fabienne Sandkuhler, Stephen Thomas, Ben Weinstein-Raun, Jan Brauner, "Thousands of AI Authors on the Future of AI," preprint in Jan. 2024, https://aiimpacts.org/wp-content/uploads/2023/04/Thousands_of_AI_authors_on_the_future_of_AI.pdf

20. Max Roser, "Artificial Intelligence Is Transforming Our World—It Is On All of Us to Make Sure That It Goes Well," Dec. 15, 2022 on *OurWorldInData.org*, https://ourworldindata.org/ai-impact

21. Benjamin Hilton, "Preventing an AI-Related Catastrophe," March 2023, on *80000hours.org*,https://80000hours.org/problem-profiles/artificial-intelligence/#neglectedness

22. "Transcript: Ezra Klein Interviews Kelsey Piper," Mar. 21, 2023 on *The Ezra Klein Show*, https://www.nytimes.com/2023/03/21/podcasts/ezra-klein-podcast-transcript-kelsey-piper.html

23. From Maureen Dowd, "Elon Musk's Billion-Dollar Crusade to Stop the A.I. Apocalypse," March 26, 2017 on *VanityFair.com*, https://www.vanityfair.com/news/2017/03/elon-musk-billion-dollar-crusade-to-stop-ai-space-x

24. Sarah Jackson, "The CEO of the Company Behind AI Chatbot ChatGPT Says the Worst-Case Scenario for Artificial Intelligence Is 'Lights Out for All of Us'," Jan. 25, 2023 on *Business Insider*, https://www.businessinsider.com/chatgpt-openai-ceo-worst-case-ai-lights-out-for-all-2023-1

25. Cade Metz, "OpenAI Unveils A.I. Agent That Can Use Websites on Its Own," Jan. 23, 2025 on *NYTimes.com*, https://www.nytimes.com/2025/01/23/technology/openai-operator-launch.html

26. Ezra Klein, "The Surprising Thing A.I. Engineers Will Tell You if You Let Them," April 16, 2023 in the *New York Times*, https://www.nytimes.com/2023/04/16/opinion/this-is-too-important-to-leave-to-microsoft-google-and-facebook.html

27. John Herrman, "How Mark Zuckerberg Led the Tech Industry Into a Metaverse Wasteland," May 10, 2023 in *New York*, https://nymag.com/intelligencer/2023/05/the-metaverse-was-a-ridiculous-idea-where-did-it-come-from.html

28. Deudney, *Dark Skies*, P. 167

29. Steven Zeitchik, "The Future of Warfare Could Be a Lot More Grisly Than Ukraine," March 11, 2022 in the *New York Times*, https://www.washingtonpost.com/technology/2022/03/11/autonomous-weapons-geneva-un/

30. James Dawes, "UN Fails to Agree on "Killer Robot" Ban as Nations Pour Billions into Autonomous Weapons Research," Dec. 20, 2021 on *TheConversation.com*, https://theconversation.com/un-fails-to-agree-on-killer-robot-ban-as-nations-pour-billions-into-autonomous-weapons-research-173616

31. Zeitchik, ibid.

32. David Lague, "U.S. and China Race to Shield Secrets from Quantum Computers," Dec. 14, 2023 on *Reuters.com*, https://www.reuters.com/investigates/special-report/us-china-tech-quantum/

33. Zach Montague, "The Race to Save Our Secrets from the Computers of the Future," Oct. 22, 2023 on *NYTimes.com*, https://www.nytimes.com/2023/10/22/us/politics/quantum-computing-encryption.html

6. "The Technical Problem Captivated Me": On Psychology And Culture

1. See, for example, Jessica Bruder, "Driven to Despair," May 14, 2018 in *New York*,https://nymag.com/intelligencer/2018/05/the-tragic-end-to-a-black-car-drivers-campaign-against-uber.html

2. "Fact Sheet: Road Safety" on *UN.org*, https://www.un.org/sites/un2.un.org/files/media_gstc/FACT_SHEET_Road_safety.pdf

3. Om Malik, "Silicon Valley Has an Empathy Bubble, Nov. 28, 2016 in *The New Yorker*, https://www.newyorker.com/business/currency/silicon-valley-has-an-empathy-vacuum

4. Quoted in "Sex, Beer, and Coding: Inside Facebook's Wild Early Days in Palo Alto," Sept. 14, 2019 on *ProductivityHub.org*, https://productivityhub.org/2019/09/14/sex-beer-and-coding-inside-facebooks-wild-early-days-in-palo-alto/

5. "Who Owns Facebook? The Definitive Who's Who Guide to Facebook Wealth," 2012 in http://whoownsfacebook.com/#Callahan

6. Quoted in Fisher, *Valley of Genius*, pp. 368-369

7. Joseph Henrich, Steven J. Heine, and Ara Norenzayan, "The Weirdest People in the World," June 15, 2010 in *Behavioral and Brain Sciences*, Volume 33 , Issue 2-3, June 2010 , pp. 61 - 83, DOI: https://doi.org/10.1017/S0140525X0999152X

8. Daub, *What Tech Calls Thinking*, p. 14

9. "2010 Census Shows America's Diversity," March 24, 2011 from *Census.gov*, https://www.census.gov/newsroom/releases/archives/2010_census/cb11-cn125.html

10. "Is Silicon Valley Tech Diversity Possible Now?" June 26, 2018 from UMass-Amherst's Center for Employment Equity, https://www.umass.edu/employmentequity/sites/default/files/CEE_Diversity+in+Silicon+Valley+Tech.pdf

11. Borsook, *Cyberselfish*, p. 41

12. "Top Companies for Women Technologists," 2020 by the AnitaB.org, https://4b7xbg26zfmr1aupi724hrym-wpengine.netdna-ssl.com/wp-content/uploads/2020/09/2020-TopCompanies-InsightReport-rFINAL.pdf

13. Ashley Bittner and Brigette Lau, "Women-led Startups Received Just 2.3& of VC Funding in 2020, Feb. 25, 20211 in *Harvard Business Review*, https://hbr.org/2021/02/women-led-startups-received-just-2-3-of-vc-funding-in-2020

14. Cook, *The Psychology of Silicon Valley*, p. 45

15. Cook, ibid. p. 54

16. Shai Danziger, Jonathan Levay, and Liora Avnaim-Pesso, "Extraneous Factors in Judicial Decisions," Apr. 11, 2022 in *Proceedings of the National Academy of Sciences of the United States of America*, PNAS.org, Vol. 108, No. 17, https://www.pnas.org/doi/full/10.1073/pnas.1018033108

17. Jack Clark, "Artificial Intelligence Has a 'Sea of Dudes' Problem," June 23, 2016 on *Bloomberg.com*, https://www.bloomberg.com/news/articles/2016-06-23/artificial-intelligence-has-a-sea-of-dudes-problem

18. Rob Reich, Mehran Sahami, and Jeremy Weinstein, *System Error: Where Big Tech Went Wrong and How We Can Reboot* (New York: HarperCollins, 2021) pp. 80-81

19. Emily Bender, Timnit Gebru, Angelina McMillan-Major, Shmargaret Shmitchell [an alias], "On the Dangers of Stochastic Parrots: Can Language Models Be Too Big?" FAccT '21: Proceedings of the 2021 ACM Conference on Fairness, Accountability, and Transparency, 2021, pages 610–623, https://s10251.pcdn.co/pdf/2021-bender-parrots.pdf

20. Cook, ibid., pp. 39-40

21. Quote from interview by James Jacoby, "The Facebook Dilemma," Feb. 28, 2018 for *Frontline*, https://www.pbs.org/wgbh/frontline/interview/sandy-parakilas/

22. Thomas A. Edison, "Improvement in Electrographic Vote-Recorder," https://patents.google.com/patent/US90646

23. "Mark Zuckerberg: Facts & Related Content," in *Brittanica.com*, https://www.britannica.com/facts/Mark-Zuckerberg#:~:text=Mark%20Zuckerberg%20was%2019%20years,(renamed%20Facebook%20in%202005).

24. Interview with the author, Feb. 27, 2022

25. Claire Hoffman, "The Battle for Facebook," Sept. 15, 2010 on *RollingStone.com*,https://www.rollingstone.com/culture/culture-news/the-battle-for-facebook-242989/

26. Alina Selyukh and Aarti Shahani, June 21, 2017 on *All Tech Considered* on *NPR.org*, https://www.npr.org/sections/alltechconsidered/2017/06/21/533791446/after-ceo-resignation-is-uber-kalanick-less-or-kalanick-free

27. Matthew Cantor, "Twitter's Been Sending Press the Poop Emoji. Why Does Musk Love It So Much?" Mar. 23, 2023 on *TheGuardian.com*, https://www.theguardian.com/technology/2023/mar/23/twitter-elon-musk-poop-emoji

28. Froma Harrop, "Why Do Investors Back Immature Males?" May 12, 2017 on *ProvidenceJournal.com*, https://www.providencejournal.com/story/opinion/2017/05/13/froma-harrop-why-do-investors-back-immature-males/21050187007/

29. Deudney, *Dark Skies*, p. 367

30. Stuart A. Thompson, "Uncensored Chatbots Provoke a Fracas Over Free Speech," July 2, 2023 on *NYTimes.com*, https://www.nytimes.com/2023/07/02/technology/ai-chatbots-misinformation-free-speech.html

31. Interview with the author, April 10, 2023

32. Kelsey Piper interviewed by Ezra Klein, March 21, 2023 on *The Ezra Klein Show*, https://www.nytimes.com/2023/03/21/podcasts/ezra-klein-podcast-transcript-kelsey-piper.html

33. Cook, ibid., 24

34. Quoted in Zack Smith, "Half of Silicon Valley Has Something You'd Call Asperger's," Feb. 21, 2011 on *IndyWeek.com*, https://indyweek.com/culture/archives/half-silicon-valley-something-call-asperger-s-interview-temple-grandin/

35. Cook, ibid., p. 26

36. Cannon, W.M., & Perry, D.K. (1966) "A Vocational Interest Scale for Computer Programmers." *Proceedings of the Fourth SIGCPR Conference on Computer Personnel Research, ACM.* pp. 61-82

37. Cook, ibid., p. 23

38. Quoted in an interview by Scott Dadich, "Barack Obama, Neural Nets, Self-Driving Cars, and the Future of the World," November 2016 in *Wired*, https://www.wired.com/2016/10/president-obama-mit-joi-ito-interview/

39. Lanier, *You Are Not a Gadget*, pp. 35-36

40. See, for example, Biz Carson, "Silicon Valley Start-ups Are Obsessed with Developing Tech to Replace Their Moms," May 10, 2015 in *Business Insider*, https://www.businessinsider.com/san-francisco-tech-startups-replacing-mom-2015-5?op=1

41. Quoted from an interview with Katy Cook on Sept. 27, 2018 in *The Psychology of Silicon Valley*

42. Interview with the author on Apr. 10, 2023

43. Om Malik, Nov. 26, 2014 on his blog *Om.co*, https://om.co/2014/11/26/technology-and-the-moral-dimension/

44. Wendy Liu, *Abolish Silicon Valley: How to Liberate Technology from Capitalism* (London: Repeater Books, 2020)

45. Liu, ibid., p. 109

46. Liu, ibid., p. 115

47. Liu, ibid., p. 112

48. Liu, ibid., pp. 131-132

49. Liu, ibid., pp. 123-124

50. Liu, ibid. P. 143

51. Interview with the author, Feb. 7, 2002

52. Maëlle Gavet, *Trampled by Unicorns: Big Tech's Empathy Problem and How to Fix It* (Hoboken: Wiley, 2021) p. 107

53. Gavet, ibid., p. 106

54. Meg Kinard, "A Comprehensive Look at DOGE's Firings and Layoffs So Far," Feb. 21, 2025 on *APNews.com*, https://apnews.com/article/doge-firings-layoffs-federal-government-workers-musk-d33cdd7872d64d2bdd8fe70c28652654

55. See https://www.youtube.com/watch?v=nkMVboRNptA

56. See https://www.youtube.com/watch?v=Sr_SPUGJTwE

7. On History, And Progress

1. Quoted in Andrew Marantz, "The Dark Side of Techno-Utopianism," Sept. 23, 2019 in *The New Yorker*, https://www.newyorker.com/magazine/2019/09/30/the-dark-side-of-techno-utopianism

2. Quoted in Charles Duhigg, "Did Uber Steal Google's Intellectual Property?" Oct. 15, 2018 in *The New Yorker*, https://www.newyorker.com/magazine/2018/10/22/did-uber-steal-googles-intellectual-property

3. James C. Scott, *Against the Grain: A Deep History of the Earliest States* (New Haven: Yale University Press, 2017) p. 9

4. Jason Crawford, "Smart, Rich, and Free," from his blog *The Roots of* Progress, July 17, 2017, https://rootsofprogress.org/smart-rich-and-free

5. Mumford, *Technics & Civilization.*, p. 108

6. Mumford, ibid., p. 109

7. David Graeber and David Wengrow, *The Dawn of Everything: A New History of Humanity* (New York: FSG, 2021), p. 29

8. Quoted in Carroll Pursell, *White Heat: People and Technology* (Berkeley: University of California Press, 1994) p. 26

9. Pursell, loc. cit.

10. Graeber & Wengrow, ibid., p. 499

11. Jared Diamond, *Guns, Germs, and Steel* (New York: W.W. Norton, paperback version, 1999) p. 215

12. Graeber & Wengrow, ibid., p. 87

13. Jared Diamond, "The Worst Mistake in the History of the Human Race" in Discovermagazine.com, May 19, 1999, https://www.discovermagazine.com/planet-earth/the-worst-mistake-in-the-history-of-the-human-race

14. Jared Diamond, *Guns, Germs, and Steel*, p. 105

15. Scott, *Against the Grain*, p. 71

16. Scott, ibid., p. 112

17. Scott, ibid., p. 107

18. Scott, ibid., pp. 18 and 20

19. R. Brian Ferguson, "War Is *Not* Part of Human Nature" in *Scientific American*, Sept. 1, 2018, https://www.scientificamerican.com/article/war-is-not-part-of-human-nature/ ; see also Nigel Barber Ph.D., "How Warlike Were Our Ancestors?" on *PsychologyToday.com*, Feb. 19, 2016, https://www.psychologytoday.com/us/blog/the-human-beast/201602/how-warlike-were-our-ancestors

20. Douglas P. Fry and Patrik Söderberg, *Journal of Aggression, Conflict, and Peace Research*, 2014 Vol. 6 No. 4, pp. 255-266

21. Diamond, ibid., pp. 196-197

22. John G. Ratey, MD and Richard Manning, *Go Wild: Eat Fat, Run Free, Be Social, and Follow Evolution's Other Rules for Total Health and Well-Being* (New York: Little, Brown Spark, 2014), p. 50

23. Ratey & Manning, ibid., p. 45
24. Diamond, ibid., p. 92
25. Scott, ibid., p. 31
26. Scott, ibid., p. 154
27. Scott, ibid., p. 155
28. Graeber & Wengrow, ibid., p. 3
29. Langdon Winner, from his 2010 foreword to Lewis Mumford's afore-cited *Technics & Civilization*, p. xii
30. Graeber and Wengrow, ibid., p. 497
31. "Time," Brittanica.com, https://www.britannica.com/science/time
32. Mumford, ibid., p. 71
33. Milton Leitenberg, "Deaths in Wars and Conflicts in the 20th Century," Cornell University Peace Studies Program Occasional Paper #29, 3rd ed., 2003, 2005, 2006, https://ecommons.cornell.edu/bitstream/handle/1813/69395/29-Leitenberg-Deaths-in-Wars-3ed.pdf, p. 1
34. Crawford, "The Most Peaceful Time in History," Sept. 21, 2017 on his *The Roots of Progress* blog, https://blog.rootsofprogress.org/the-most-peaceful-time-in-history
35. Steven Johnson, *Wonderland: How Play Made the Modern World* (New York: Riverhead Books, 2016) p. 36
36. Wayne Grady, *Technology* (Toronto: Groundwood Books, 2010) p. 111
37. Graeber & Wengrow, ibid., pp. 71-73
38. Graeber & Wengrow, ibid., p. 45
39. Graeber & Wengrow, ibid., p. 41
40. Graeber & Wengrow, ibid., p. 39
41. Graeber & Wengrow, ibid., pp. 37-56
42. Graeber & Wengrow, ibid., see their chapter "Wicked Liberty."
43. Ratey & Manning, ibid., p. 152

8. How We Think Of Tech

1. Rebecca Solnit, "Donald Trump's Power Is Fading: Trumpism Is the Clear and Present Danger Now," March 8, 2022 in the *Guardian*, https://www.theguardian.com/commentisfree/2022/mar/08/donald-trump-power-trumpism-danger-now
2. Kevin Kelly, *The Inevitable: Understanding the 12 Technological Forces That Will Shape Our Future* (New York, Penguin Books, 2016) p. 2
3. Kevin Kelly, *What Technology Wants* (New York: Penguin Books, 2011)
4. Mumford, *Technics & Civilization*, p. 6
5. John Harris, "There Was All Sorts of Toxic Behaviour': Timnit Gebru on Her Sacking by Google, AI's Dangers and Big Tech's Biases," May 22, 2023 on *TheGuardian.com*,https://www.theguardian.com/lifeandstyle/2023/may/

22/there-was-all-sorts-of-toxic-behaviour-timnit-gebru-on-her-sacking-by-google-ais-dangers-and-big-techs-biases

6. William Butler Yeats, "Responsibilities," *The Collected Works of W.B. Yeats, Volume I: The Poems: Revised Second Edition,* (New York: Scribner, 2010) p. 102

7. Daniel Chandler, "Technological or Media Determinism," Sept. 18, 1995," http://visual-memory.co.uk/daniel//Documents/tecdet/tdet12.html

8. Kelly, *The Inevitable*, p. 5

9. Langdon Winner, *Autonomous Technology: Technics-out-of-Control as a Theme in Political Thought,* (Cambridge: MIT Press, 1978) p. 99

10. Wendell Wallach, *A Dangerous Master: How to Keep Technology from Slipping Beyond Our* Control (New York: Basic Books, 2015) p. 71

11. Lanier, *You Are Not a Gadget*, pp. 65 and 67

12. Kelsey Piper interviewed by Ezra Klein, March 21, 2023 on *The Ezra Klein Show,* https://www.nytimes.com/2023/03/21/podcasts/ezra-klein-podcast-transcript-kelsey-piper.html

13. In *OxfordReference.com*,https://www.oxfordreference.com/display/10.1093/acref/9780191826719.001.0001/q-oro-ed4-00007996

14. Ezra Klein, "The Surprising Thing A.I. Engineers Will Tell You if You Let Them," Apr. 16, 2023 on *NYTimes.com,* https://www.nytimes.com/2023/04/16/opinion/this-is-too-important-to-leave-to-microsoft-google-and-face book.html

15. See for example, Tristan Harris and Aza Raskin in conversation, "The A.I. Dilemma," March 24, 2023 on the *Your Undivided Attention* podcast, https://www.humanetech.com/podcast/the-ai-dilemma

16. Ursula Franklin, *The Real World of Technology* (House of Anansi Press, 2004; first published 1990) p. 25

17. Daniel De Visé, "A Record Share of Americans Is Living Alone," July 10, 2023 on *TheHill.com,* https://thehill.com/policy/healthcare/4085828-a-record-share-of-americans-are-living-alone/#:~:text=While%20nearly%2030%20percent%20of,of%20Americans%20live%20in%20them.

18. See, for example, Dan Vahaba, "Could AI-powered Robot 'Companions' Combat Human Loneliness," July 12, 2023 on *Today.Duke.Edu,* https://today.duke.edu/2023/07/could-ai-powered-robot-companions-combat-human-loneliness

19. Richard Weissbourd, Milena Batanova, Virginia Lovison, and Eric Torres, "Loneliness in America: How the Pandemic Has Deepened an Epidemic of Loneliness and What We Can Do About It," Feb. 2021 from the Making Caring Common Project, Harvard Graduate School of Education, https://static1.squarespace.com/static/5b7c56e255b02c683659fe43/t/6021776bd d04957c4557c212/1612805995893/Loneliness+in+America+2021_02_08_FINAL.pdf

20. David Wallace Wells, "Talking with Kevin Kelly About the Future of Tech and Our Species Identity Crisis," June 15, 2016 in *New York*, https://nymag.

com/intelligencer/2016/06/talking-with-wired-founding-editor-kevin-kelly-about-our-species-identity-crisis.html

21. Marc Andreessen, "Why AI Will Save the World," June 6, 2023 on *Marc Andreessen Substack*, https://pmarca.substack.com/p/why-ai-will-save-the-world

22. Ted Chiang, "Will A.I. Become the New McKinsey?" May 4, 2023 in *The New Yorker*, https://www.newyorker.com/science/annals-of-artificial-intelligence/will-ai-become-the-new-mckinsey

9. Up to the Stars: Space Colonization and Messaging E.T.s

1. https://tomlehrersongs.com/wp-content/uploads/2018/12/wernher-von-braun.pdf

2. See, for example, Mara Gordon, "What's Your Purpose? Finding Meaning in Life is Linked to Health," May 25, 2019 on *Npr.org*, https://www.npr.org/sections/health-shots/2019/05/25/726695968/whats-your-purpose-finding-a-sense-of-meaning-in-life-is-linked-to-health

3. Natalie Goldberg, *The Great Spring: Writing, Zen, and This Zigzag Life* (Boulder: Shambhala Publications, 2016) p. 195

4. William Shakespeare, *Henry V*, Act IV, Scene 3.

5. Antonio Garcia Martinez, *Chaos Monkeys: Obscene Fortune and Random Failure in Silicon Valley* (New York: HarperCollins, 2016, 2018) p. 283

6. Martinez, ibid., p. 285

7. Paulina Borsook, *Cyberselfish*, p. 133

8. Cook, *The Psychology of Silicon Valley*, p. 29

9. Cook, ibid. p. 106

10. Kara Swisher, "Is the Problem Facebook? Or the Internet?" Nov. 4, 2021 in the *New York Times*, https://www.nytimes.com/2021/11/04/opinion/sway-kara-swisher-casey-newton.html?showTranscript=1

11. See, for example, Nathaniel Rich, "Can Humans Endure the Psychological Torment of Mars?" Feb. 25 in the *New York Times*, https://www.nytimes.com/2024/02/25/magazine/mars-isolation-experiment.html

12. Francis A. Cucinotta and Eliedonna Cacao, "Non-Targeted Effects Models Predict Significantly Higher Mars Mission Cancer Risk Than Targeted Effects Models," May 12, 2017 in *Scientific Reports* 7, Article Number 1832 (2017), https://www.nature.com/articles/s41598-017-02087-3

13. Joshua Rothman, "Can Science Fiction Wake Us Up to Our Climate Reality," Jan. 24, 2022 in *The New Yorker*, https://www.newyorker.com/magazine/2022/01/31/can-science-fiction-wake-us-up-to-our-climate-reality-kim-stanley-robinson

14. Daniel Deudney, *Dark Skies: Space Expansionism, Planetary Geopolitics, & the Ends of Humanity* (New York: Oxford University Press, 2020) P. 368

Notes

15. Eric Lipton, "New Star Wars Plan: Pentagon Rushes to Counter Threats in Orbit," May 17, 2024 in the *New York Times*, https://www.nytimes.com/2024/05/17/us/politics/pentagon-space-military-russia-china.html?searchResultPosition=3

16. Simone McCarthy, "America's Military Has the Edge in Space. China and Russia Are in a Counterspace Race to Disrupt It," May 27, 2024 on *CNN.com*, https://www.cnn.com/2024/05/27/china/counterspace-us-china-russia-intl-hnk-scn/index.html

17. Deudney, ibid., p. 265

18. Interview with the author, Feb. 4, 2022

19. Mumford, *Technics & Civilization*, p. 74

20. Kate Crawford, *Atlas of AI* (New Haven: Yale University Press Books, 2021) p. 26

21. Interview with the author, Feb. 4, 2022

22. Interview with the author, Feb. 4, 2022

23. Interview with the author, Feb. 4, 2022

24. Deudney, ibid., P. 7

25. Interview with the author, Feb. 4, 2022

26. Deudney, ibid., pp. 13-14

27. Deudney, ibid., p. 219

28. Walter Isaacson, "'How Am I in this War?': The Untold Story of Elon Musk's Support for Ukraine," Sept. 7, 2023 on *WashingtonPost.com, https://www.washingtonpost.com/opinions/2023/09/07/elon-musk-starlink-ukraine-russia-invasion/#*

29. Adam Satariano, Scott Reinhard, Cade Metz, Sheera Frenkel, and Malika Khurana, "Elon Musk's Unmatched Power in the Stars," July 28, 2023 on *NYTimes.com*, https://www.nytimes.com/interactive/2023/07/28/business/starlink.html#

30. Deudney, ibid., P. 24

31. Deudney, ibid., p. 46

32. Interview with the author, Feb. 4, 2022

33. Wilson said this in a debate at the Harvard Museum of Natural History, Cambridge, Massachusetts on Sept. 9, 2009, cited in *Oxford Essential* Quotations, 4th edition published online by Oxford University Press in 2016, https://www.oxfordreference.com/view/10.1093/acref/9780191826719.001.0001/q-oro-ed4-00016553

34. Deudney, ibid., p. 357

35. This information about sending messages out into the Universe comes from Steven Johnson, "Greetings, E.T. (Please Don't Murder Us.)," June 28, 2017 in the *New York Times Magazine*, https://www.nytimes.com/2017/06/28/magazine/greetings-et-please-dont-murder-us.html

36. Johnson, loc. cit.

37. Johnson, loc. cit.

38. "How to Send a Message to Another Planet," Nov. 16, 2017 in the *Economist*, https://www.economist.com/science-and-technology/2017/11/16/how-to-send-a-message-to-another-planet

39. Steven Johnson, loc. cit.

40. Jeremy Kahn, "Just How Shallow Is the Artificial Intelligence Talent Pool?" Feb. 7, 2018 on *IndustryWeek.com*, https://www.industryweek.com/technology-and-iiot/emerging-technologies/article/22025095/just-how-shallow-is-the-artificial-intelligence-talent-pool

10. On Ideology: A Very Strange Icon

1. Peter Thiel, "The Education of a Libertarian," Apr. 13, 2009 on *Cato Unbound,* https://www.cato-unbound.org/2009/04/13/peter-thiel/education-libertarian/

2. Nick Bilton, "Silicon Valley's Most Disturbing Obsession," Oct. 5, 2016 on *Vanity Fair.com*, https://www.vanityfair.com/news/2016/10/silicon-valley-ayn-rand-obsession

3. Lisa Duggan, *Mean Girl: Ayn Rand and the Culture of Greed* (Oakland: University of California Press, 2019) pp. xiii-xiv

4. Duggan, ibid., p. 11

5. Johann Hari, "How Ayn Rand Became an American Icon," Nov. 2, 2009 on *Slate.com*, https://slate.com/culture/2009/11/two-biographies-of-ayn-rand.html

6. Duggan, ibid., p. 39

7. Duggan, ibid., p. 68

8. Duggan, ibid., p. 38

9. Duggan, ibid., p. 55

10. Duggan, ibid., pp. 40-41

11. Duggan, ibid., p. 74

12. Duggan, ibid., p. 69

13. Duggan, ibid., p. 11

14. Duggan, ibid., p. 75

15. Ben Norton, quoting from the transcript of a Q&A session with Rand after a speech she made at the U.S. Military Academy at West Point on March 6, 1974, "Libertarian Superstar Ayn Rand Defended Native American Genocide: 'Racism Didn't Exist in this Country Until the Liberals Brought It Up'," Oct. 14, 2015 on *Salon.com*, https://www.salon.com/2015/10/14/libertarian_superstar_ayn_rand_defended_genocide_of_savage_native_americans/

16. Rund Abdelfatah et. al., "The Monster of We," *NPR.org* transcript, Dec. 16, 2021, https://www.npr.org/transcripts/1064517537

17. "When Ayn Rand Collected Social Security & Medicare, After Years of Opposing Benefit Programs," Dec. 27, 2016 on *OpenCulture.com*, https://

www.openculture.com/2016/12/when-ayn-rand-collected-social-security-medicare.html

18. Liu, *Abolish Silicon Valley*, p. 23
19. Abdelfatah, loc. cit.
20. Abdelfatah, loc. cit.
21. Bilton, ibid.
22. Jonathan Freedland, "The New Age of Ayn Rand: How She Won over Trump and Silicon Valley," April 10, 2017 on *TheGuardian.com*, https://www.theguardian.com/books/2017/apr/10/new-age-ayn-rand-conquered-trump-white-house-silicon-valley
23. For an example of this type of thinking, see Mike Masnick, "We're Going Down a Very Dangerous Path With AI Regulation, Despite Better Options," July 7, 2023 on *TechDirt.com*, https://www.techdirt.com/2023/07/07/were-going-down-a-very-dangerous-path-with-ai-regulation-despite-better-options/
24. Quoted in an interview by Scott Dadich, "Barack Obama, Neural Nets, Self-Driving Cars, and the Future of the World," November 2016 in *Wired* magazine,https://www.wired.com/2016/10/president-obama-mit-joi-ito-interview/
25. Borsook, *Cyberselfish.*, pp. 152-153
26. "John Perry Barlow, "A Declaration of the Independence of Cyberspace," Jun. 1, 1996 in *Wired*, https://www.wired.com/1996/06/declaration-independence-cyberspace/
27. P.E. Mosokowitz, "The Internet Is Destroying Our Brains, But We Can't Quit. It's a Factory We're Forced to Work in Without Any Pay," June 19, 2021 in *BusinessInsider.com*, https://www.businessinsider.in/tech/news/the-internet-is-destroying-our-brains-but-we-cant-quit-its-a-factory-were-forced-to-work-in-without-any-pay-/articleshow/83666612.cms
28. Borsook, ibid., pp. 122-129
29. "Fairness Doctrine," in Brittanica.com, https://www.britannica.com/topic/Fairness-Doctrine#ref1248880
30. Wallach, *A Dangerous Master*, p. 85
31. Langdon Winner, *The Whale and the Reactor: A Search for Limits in an Age of High Technology* (Chicago: The University of Chicago Press, 1989), p. 81
32. Milton Friedman, "A Friedman Doctrine—The Social Responsibility of Business Is to Increase Its Profits," Sept. 13, 1970 in the *New York Times*, https://www.nytimes.com/1970/09/13/archives/a-friedman-doctrine-the-social-responsibility-of-business-is-to.html
33. Jonathan Taplin, *Move Fast and Break Things: How Facebook, Google, and Amazon Cornered Culture and Undermined Democracy* (Back Bay Books: 2017) p. 193
34. Darrell M. West, "It Is Time to Restore the US Office of Technology Assessment," Feb. 10, 2021 in a Brookings Institute Report, https://www.brookings.edu/research/it-is-time-to-restore-the-us-office-of-technology-assessment/
35. Zuboff, *The Age of Surveillance Capitalism* p. 103

36. Thiel, "The Education of a Libertarian," ibid.

37. J. Storrs Hall, *Where Is My Flying Car?* (San Francisco: Stripe Press, 20210)

38. Hall, ibid., p. 116

39. Borsook, *Cyberselfish.*, p. 93

40. William Paterson, *Robert A. Heinlein: 1948-1988, The Man Who Learned Better* (New York: Tom Doherty Associates, 2014) p. 389

41. Quoted in Liz Jacobs, "GPS, Lithium Batteries, the Internet, Cellular Technology, Airbags: A Q&A About How Governments Often Fuel Innovation," Oct. 28, 2013 on *TEDBlog*, https://blog.ted.com/qa-mariana-mazzucato-governments-often-fuel-innovation/

42. Jacobs, loc. cit.

43. Borsook, ibid., p. 4

44. Maggie Severns, Matt Stiles, and Alex Leeds Matthews, "Elon Musk Hates Government Subsidies. His Companies Love Them," April 30, 2022 on *Grid.news*, https://www.grid.news/story/technology/2022/04/30/elon-musk-hates-the-government-his-companies-love-it/

45. Severns, Stiles, and Matthews, loc. cit.

46. Desmond Butler, Trisha Thadani, Emmanuel Martinez, Aaron Gregg, Luis Melgar, Jonathan O'Connell, and Dan Keating, "Elon Musk's Business Empire Is Built on $38 Billion in Government Funding," Feb. 26, 2025 on *WashingtonPost.com*, https://www.washingtonpost.com/technology/interactive/2025/elon-musk-business-government-contracts-funding/

47. Jill Lepore, "The Failed Ideas That Drive Elon Musk," April 4, 2025 on *NYTimes.com*, https://www.nytimes.com/2025/04/04/opinion/elon-musk-doge-technocracy.html

48. David Broockman, Gregory Ferenstein, and Neil Malhotra, "Predispositions and the Political Behavior of American Economic Elites: Evidence from Technology Entrepreneurs," *American Journal of Political Science*, Jan. 2019, Vol. 63, Issue 1, pp. 212-233.

49. Mary Clare Jalonick and Barbara Ortutay, "Zuckerberg: Regulation 'Inevitable'for Social Media Firms," Apr. 12, 2018 on *APNews.com*, https://apnews.com/article/04076e9451481477cb9e08a5383528b15

50. Cecilia Kang, "OpenAI's Sam Altman Urges A.I. Regulation in Senate Hearing," May 16, 2023 on *NYTimes.com*, https://www.nytimes.com/2023/05/16/technology/openai-altman-artificial-intelligence-regulation.html#:~:text=In%20his%20first%20testimony%20before,over%20A.I.%27s%20potential%20harms.

51. Cecilia Kang, "Emboldened by Trump, A.I. Companies Lobby for Fewer Rules," Mar. 24, 2025 on *NYTimes.com*, https://www.nytimes.com/2025/03/24/technology/trump-ai-regulation.html

52. Antonio Pequeno IV, "JD Vance and Peter Thiel: What to Know About the Relationship Between Trump's VP Pick and the Billionaire," July 16, 2024 on *Forbes.com*, https://www.forbes.com/sites/antoniopequenoiv/2024/07/16/

jd-vance-and-peter-thiel-what-to-know-about-the-relationship-between-trumps-vp-pick-and-the-billionaire/

53. "JD Vance at AI Action Summit in France," Feb. 11, 2025 on *YouTube.com*, https://www.youtube.com/watch?v=n_0c6F4ykSs

11. The Elephants In The Room: Venture Capitalists, Monopolists, And Hyper-Growth

1. McNamee, *Zucked*, p. 235
2. Douglas Rushkoff, *Throwing Rocks at the Google Bus: How Growth Became the Enemy of Prosperity* (New York: Portfolio/Penguin, 2016) p. 16
3. Interview with the author, ibid.
4. Taplin, *Move Fast and Break Things*, p. 201
5. Rushkoff, ibid., p. 191
6. 8 Tarnoff & Weigel, *Voices from the Valley*, p. 29
7. Martinez, *Chaos Monkeys*, p. 103
8. Fisher, *Valley of Genius*, p. 368
9. Lanier, *You Are Not a Gadget*, p. 16
10. Arun Rao, "A History of Silicon Valley: The Greatest Creation of Wealth in the History of the Planet," (Omniware, 2011) https://www.scaruffi.com/svhistory/arun6.html
11. See, for example, Willard F. Mueller, "Antitrust in the Reagan Administration," 1984 in *Revue Francaise d'Etudes Americaines*, 21-22, pp. 427-434, https://www.persee.fr/doc/rfea_0397-7870_1984_num_21_1_1187
12. *Statcounter GlobalStats*, Nov. 2020-Nov. 2021, https://gs.statcounter.com/search-engine-market-share
13. "Largest American Companies by Market Capitalization," https://companiesmarketcap.com/usa/largest-companies-in-the-usa-by-market-cap/ accessed on Jan. 14, 2024
14. Rushkoff, ibid., p. 71
15. Wu, *The Master Switch*, p. 49
16. Cook, *The Psychology of Silicon Valley*, p. 276
17. Wu, ibid., p. 107
18. Yiren Lu, "A Week With the Wild Children of the A.I. Boom," May 31, 2023 in the *New York Times Magazine*, https://www.nytimes.com/2023/05/31/magazine/ai-start-up-accelerator-san-francisco.html
19. Cade Metz, "In the Age of A.I., Tech's Little Guys Need Big Friends," July 5, 2023 on *NYTimes.com*, https://www.nytimes.com/2023/07/05/business/artificial-intelligence-power-data-centers.html
20. Doctorow, *How to Destroy Surveillance Capitalism*, p. 90
21. Wu, ibid., p. 107
22. Wu, ibid., p. 127

23. McNamee, *Zucked*, p.141

24. Doctorow, ibid., p. 58

25. Gavet, *Trampled by Unicorns*, p. 72

26. Gavet, ibid., pp. 72-73

27. Interview by Noah Kulwin, "Facebook is a Fundamentally Addictive Product," April 2018 in *New York* magazine, https://nymag.com/intelligencer/2018/04/sandy-parakilas-former-facebook-employee-interview.html

28. Interview with the author, April 10, 2023

29. Rushkoff, ibid., pp. 9-10

30. Rushkoff, ibid., p. 15

31. Edward Abbey, from *The Journey Home: Some Words in the Defense of the American West*, quoted in Martinez, ibid., p. 373

32. McNamee, ibid., p. 292

33. Rushkoff, ibid. p.118

34. Zuboff, *The Age of Surveillance Capitalism*, p. 101

35. Max Chafkin, "Peter Thiel's Origin Story," Sept. 20, 2021 in *New York* magazine,https://nymag.com/intelligencer/article/peter-thiel-silicon-valley-contrarian-max-chafkin.html

36. Tavis McGinn, former Facebook pollster, quoted in "Facebook Hired a Full-Time Pollster to Monitor Zuckerberg's Approval Ratings: Why Tavis McGinn Quit After Six Months" by Casey Newton on *TheVerge.com* on Feb. 6, 2018, https://www.theverge.com/2018/2/6/16976328/facebook-mark-zuckerberg-pollster-tavis-mcginn-honest-data

37. As quoted in Cook, *The Psychology of Silicon Valley*, p. 257

38. Quoted anonymously in *Voices from the Valley*, edited by Ben Tarnoff and Moira Weigel (New York: FSG Originals X Logic, 2020) pp. 31-32

39. Georgia Wells, Jeff Horwitz, and Deepa Seetharaman, "Facebook Knows Instagram Is Toxic for Teen Girls, Company Documents Show," Sept. 14, 2021 in the *Wall Street Journal*, https://www.wsj.com/articles/facebook-knows-instagram-is-toxic-for-teen-girls-company-documents-show-11631620739

III. SOLUTIONS

1. Fisher, *Valley of Genius*, p. 361

2. Fisher, loc. cit.

3. Adam Hayes, "The Unexpected Consequences of Self-Driving Cars," May 15, 2021 on *Investopedia.com*, https://www.investopedia.com/articles/investing/090215/unintended-consequences-selfdriving-cars.asp

12. Science Fiction: Sounds Of Thunder

1. Burke, *Connections.*, p. 75
2. See, for example, Arielle Contreras, "8 Pieces of Modern Technology That Science Fiction Predicted...or Invented," Oct. 16, 2017 on *ElecticLiterature.com*, https://electricliterature.com/8-pieces-of-modern-technology-that-science-fiction-predictedor-invented/
3. See, for example, Jeff Butts, "The World's First Metaverse Movies Star...You," Feb. 7, 2022 on *MacObserver.com*, https://www.macobserver.com/news/the-worlds-first-metaverse-movies-star-you/
4. Stephen Dubner, "Philip Rosedale Answers Your Second Life Questions," Dec. 13, 2007 on *Freakonomics.com*, https://freakonomics.com/2007/12/philip-rosedale-answers-your-second-life-questions/?hp
5. Neal Stephenson, *Snow Crash*, (2017 Del Rey paperback reprint of the 1992 book published by Bantam Spectra) p. 47
6. Interview with the author, Mar. 11, 2022
7. Kim Stanley Robinson, *The Ministry for the Future* (New York: Orbit Books, 2021) P. 349
8. See, for example, David E. Sanger, Eric Lipton, Eileen Sullivan and Michael Crowley, "Before Virus Outbreak, a Cascade of Warnings Went Unheeded," Mar. 19, 2020 in the *New York Times*, https://www.nytimes.com/2020/03/19/us/politics/trump-coronavirus-outbreak.html
9. Steven Johnson, *Farsighted: How We Make the Decisions That Matter the Most* (Great Britain: John Murray Publishers, 2018) p. 141
10. Steven Johnson, "Beyond the Bitcoin Bubble," the *New York Times Magazine*, Jan. 16, 2018, https://www.nytimes.com/2018/01/16/magazine/beyond-the-bitcoin-bubble.html
11. Franklin, *The Real World of Technology*, p. 95
12. See, for example, David E. Rosenbaum, "For the Highway Lobby, a Rocky Road Ahead," April 2, 1972 in the *New York Times*, https://www.nytimes.com/1972/04/02/archives/for-the-highway-lobby-a-rocky-road-ahead.html
13. Quoted anonymously in Tarnoff & Weigel, *Voices from the Valley*, p. 101

13. A Predictor's Toolbox

1. As quoted in J. Peter Scoblic, "We Can't Prevent Tomorrow's Catastrophes Unless We Imagine Them Today," Mar. 28, 2021 in the *New York Times*, https://www.washingtonpost.com/outlook/2021/03/18/future-forecasting-strategic-planning/
2. Norman Dalkey and Olaf Helmer, "An Experimental Application of the DELPHI Method to the Use of Experts," April 1, 1963 in *Management Science*, https://doi.org/10.1287/mnsc.9.3.458
3. Steven Johnson, *Farsighted*, p. 113

Notes

4. Pierre Wack, "Scenarios: Shooting the Rapids," Nov. 1985 in *Harvard Business Review*, https://hbr.org/1985/11/scenarios-shooting-the-rapids

5. Wack, loc. cit.

6. Johnson, *Farsighted*, pp. 146-147

7. Sharon Goldman, "Legions of DEF CON Hackers Will Attack Generative AI Models," Aug. 10, 2023 in *VentureBeat.com*, https://venturebeat.com/ai/legions-of-defcon-hackers-will-attack-generative-ai-models/

8. See, for example, Alice McCarthy, "Exxon Disputed Climate Findings for Years. Its Scientists Knew Better," Jan. 12, 2023 in the *Harvard Gazette*, https://news.harvard.edu/gazette/story/2023/01/harvard-led-analysis-finds-exxonmobil-internal-research-accurately-predicted-climate-change/

9. See, for example, "Big Tobacco Knew Radioactive Particles in Cigarettes Posed Cancer Risks But Kept Quiet," Sept. 28, 2011 on *UCLA-Health.org*,https://www.uclahealth.org/news/release/big-tobacco-knew-radioactive-particles-in-cigarettes

10. David Banta, "What Is Technology Assessment?" July 1, 2009 in *The International Journal of Technology Assessment and Health Care*, pp. 7-9, https://www.cambridge.org/core/journals/international-journal-of-technology-assessment-in-health-care/article/what-is-technology-assessment/F2CE6903EE02296499AC0AC24192454E

11. Langdon Winner, *The Whale and the Reactor: A Search for Limits in an Age of High Technology* (The University of Chicago Press, Chicago, 1986) p. 10

12. Winner, loc. cit.

13. Winner, loc. cit.

14. Interview with the author, Feb. 21, 2022

15. Gary Klein, "Performing a Project Premortem," Sept. 2007 in the *Harvard Business Review*, https://hbr.org/2007/09/performing-a-project-premortem

16. Interview with the author, Feb. 21, 2022

17. Interview with the author, Feb. 21, 2022

18. Interview with the author, Feb. 21, 2022

19. Interview with the author, Feb. 21, 2022

20. Interview with the author, Feb. 21, 2022

21. Klein, "Performing a Project Premortem," loc. cit.

22. Philip Tetlock and Dan Gardner, *Superforecasting: The Art and Science of Prediction* (New York: Broadway Books, 2015), p. 4

23. Tetlock and Gardner, ibid., p. 72

24. Tetlock and Gardner, ibid., p. 138

25. J. Peter Scoblic and Philip E. Tetlock "We Didn't See Donald Trump Coming. But We Could Have," Feb. 12, 2016 in the *New York Times*, https://www.washingtonpost.com/opinions/we-didnt-see-donald-trump-coming-but-we-could-have/2016/02/12/46ece26a-d0db-11e5-abc9-

26. Tetlock and Gardner, ibid., p. 79

27. Julia Galef interview with Tetlock, "Predicting the Future Is Possible. These 'Superforecasters' Know How," Dec. 3, 2021 in the *New York Times*, https://

www.nytimes.com/2021/12/03/opinion/ezra-klein-podcast-philip-tetlock.
html?showTranscript=1

28. Tetlock and Gardner, ibid., pp. 110-111

29. Tetlock and Gardner, ibid., p. 162

30. Irving Janis, *Victims of Groupthink: A Psychological Study of Foreign-Policy Decisions and Fiascoes* (Boston: Houghton-Mifflin, 1972) p. 20

31. Tetlock and Gardner, ibid., p. 254

32. Interview with the author, Mar. 11, 2022

33. Tetlock and Gardner, ibid., p. 123

34. Tetlock and Gardner, ibid., p. 279

35. Galef interview with Tetlock, ibid.

36. Galef interview with Tetlock, ibid.

37. Wallach, *A Dangerous Master*, p. 38

38. Interview with the author, Mar. 11, 2022

39. Tetlock and Gardner, ibid., p. 153

40. Tetlock and Gardner, ibid., p. 202

41. Tetlock and Gardner, ibid., p. 131

42. Steven Johnson, "Facing a Tough Decision: Borrow From Psychology, Business, and the Military to See Past Your Blind Spots," Sept. 28, 2018 on *Ideas.TED.com*,https://ideas.ted.com/facing-a-tough-decision-borrow-from-psychology-business-and-the-military-to-see-past-your-blind-spots/

43. Johnson, *Farsighted*, p. 68

44. Kahneman, *Thinking, Fast and Slow* (New York: Farrar, Straus, and Giroux, 2011)

45. Interview with the author, Feb. 21, 2022

46. Johnson, *Farsighted*, p. 144

47. Johnson, ibid, p. 202

48. Margaret Stewart. "Breadth & Depth: Why I'm Optimistic about Facebook's Responsible Innovation Efforts," June 17, 2021 on *Tech.Facebook.com*, https://tech.facebook.com/ideas/2021/6/responsible-innovation/

49. Jeff Horwitz, "Facebook Parent Meta Platforms Cuts Responsible Innovation Team," Sept. 8, 2022 on *WSJ.com*, https://www.wsj.com/articles/facebook-parent-meta-platforms-cuts-responsible-innovation-team-11662658423

14. Better Questions: At The Center Of The Wheel

1. Authman Apatira, "Ethics and The Unintended Consequences of Technology" June 1, 2018 on *CodingDojo.com*, https://www.codingdojo.com/blog/ethics-unintended-consequences-technology

2. Nitin Nohria and Hemant Taneja, "Managing the Unintended Consequences of Your Innovations," January 19, 2021 in *Harvard Business Review*,

https://hbr.org/2021/01/managing-the-unintended-consequences-of-your-innovations

3. L.M. Sacasas, "The Questions Concerning Society," June 4, 2021 on *The Convivial Society* blog, https://theconvivialsociety.substack.com/p/the-questions-concerning-technology?s=r

15. A Shared Fight

1. Zuboff, *The Age of Surveillance Capitalism*, p. 485
2. Scott, *Against the Grain*, p. xi
3. Diamond, *Guns, Germs, and Steel*, p. 105
4. Scott, ibid., p. 8
5. Graeber & Wengrow, *The Dawn of Everything*, p. 89
6. Graeber & Wengrow, ibid., pp. 127-128
7. Graeber & Wengrow, ibid., p. 517
8. Graeber & Wengrow, ibid., p. 4
9. Graeber & Wengrow, ibid., p. 235
10. Graeber & Wengrow, ibid., p. 382
11. Postman, *Technolopy*, p. 23
12. Mumford, *Technics & Civilization*, P. 85
13. Howard W. French, *Everything Under the Heavens: How the Past Helps Shape China's Push for Global Power* (New York: Knopf, 2017) p. 108; see also Jim Edwards, "500 Years Ago, China Destroyed Its World-Dominating Navy Because Its Political Elite Was Afraid of Free Trade," March 5, 2017 in *The Independent*, https://www.independent.co.uk/news/world/americas/500-years-ago-china-destroyed-its-worlddominating-navy-because-its-political-elite-was-afraid-of-free-trade-a7612276.html
14. Postman, ibid., p. 43
15. Calestous Juma, "When People Become Innovation's Greatest Threats," July 17,2016,https://theconversation.com/when-people-become-innovations-greatest-threats-62502
16. Jeff Haden, "Mark Cuban: The World's First Trillionaire Is Learning 1 Skill and Discovering How to Use It in Now Unimaginable Ways," Nov. 2, 2020 in *Inc.com*, https://www.inc.com/jeff-haden/mark-cuban-worlds-first-trillionaire-is-learning-1-skill-discovering-how-to-use-it-in-now-unimaginable-ways.html
17. Adi Gaskell, "Where Might Protests Against The 4th Industrial Revolution Originate?" Dec. 2, 2020 in *Forbes.com*, https://www.forbes.com/sites/adigaskell/2020/12/02/where-might-protests-against-the-4th-industrial-revolution-originate/?sh=5352a0bd2ecd
18. Crawford, *Atlas of AI*, p. 81
19. Nina Shapiro, "Under Pressure, Afraid to Take Bathroom Breaks? Inside Amazon's Fast-paced Warehouse World," July 2, 2018 in the *Seattle-*

Times.com, https://www.seattletimes.com/business/amazon/under-pressure-afraid-to-take-bathroom-breaks-inside-amazons-fast-paced-warehouse-world/

20. Lanier, *You Are Not a Gadget*, pp. 54-55
21. Erica Fink, "Google Glass Wearers Can Steal Your Password," July 9, 2014 on *Money.cnn.com*, https://money.cnn.com/2014/07/07/technology/security/google-glass-password-hack/index.html?iid=HP_River
22. Ryan Mac and Craig Silverman, "Hurting People at Scale: Facebook's Employees Reckon with the Social Network They've Built," July 23, 2020 on *BuzzFeedNews.com*, https://www.buzzfeednews.com/article/ryanmac/facebook-employee-leaks-show-they-feel-betrayed?bfsource=relatedmanual
23. Quoted in "The Organic Side, to Me, Is Scarier Than the Ad Side" by Noah Kulwin in *New York*, April 2018, https://nymag.com/intelligencer/2018/04/antonio-garcia-martinez-former-facebook-employee-interview.html
24. McNamee, *Zucked*, p. 236
25. Tarnoff and Weigel, *Voices from the Valley*, p. 96
26. Cook, *The Psychology of Silicon Valley*, p. 265
27. Rob Reich, Mehran Sahami, and Jeremy Weinstein, *System Error: Where Big Tech Went Wrong and How We Can Reboot* (New York: HarperCollins, 2021) p. 17
28. Crawford, *Atlas of AI*, p. 190
29. Cook, ibid., pp. 263-264
30. Karen Hao, "The Two-Year Fight to Stop Amazon from Selling Face Recognition to the Police," June 12, 2020 on *TechnologyReview.com*, https://www.technologyreview.com/2020/06/12/1003482/amazon-stopped-selling-police-face-recognition-fight/
31. Greg Bensinger, "How Illinois Is Winning in the Fight Against Big Tech," May 30, 2022 in the *New York Times*, https://www.nytimes.com/2022/05/30/opinion/illinois-biometric-data-privacy.html
32. Karen Weise, "Amazon Indefinitely Extends a Moratorium on the Police Use of Its Facial Recognition Software," Aug. 1, 2021 in the *New York Times*, https://www.nytimes.com/2021/05/18/business/amazon-police-facial-recognition.html
33. Paresh Dave, "U.S. Cities Are Backing Off Banning Facial Recognition as Crime Rises," May 12, 2022 on *Reuters.com*, https://www.reuters.com/world/us/us-cities-are-backing-off-banning-facial-recognition-crime-rises-2022-05-12/
34. Elizabeth Dwoskin and Drew Harwell, "Facebook Is Ending Use of Facial Recognition Software, Deleting Data on More Than a Billion People," Nov. 2, 2021 in the *Washington Post*, https://www.washingtonpost.com/technology/2021/11/02/facebook-ends-facial-recognition/
35. Noah Kulwin, "The Internet Apologizes for Violating and Hijacking Our Attention," April 16, 2018 in *New York*, https://nymag.com/intelligencer/2018/04/an-apology-for-the-internet-from-the-people-who-built-it.html
36. Kulwin, loc. cit.

Notes

37. Interview with the author, March 25, 2022

38. David Kushner, "The Flight of the Birdman: Flappy Bird Creator Dong Nguyen Speaks Out," March 11, 2014 on *RollingStone.com*, https://www.rollingstone.com/culture/culture-news/the-flight-of-the-birdman-flappy-bird-creator-dong-nguyen-speaks-out-112457/

39. Feb. 8, 2014, https://twitter.com/dongatory/status/432227971173068800

40. Lan Anh Nguyen, "Exclusive: Flappy Bird Creator Dong Nguyen Says App 'Gone Forever' Because It Was 'An Addictive Product'," Feb. 11, 2014 on *Forbes.com*, https://www.forbes.com/sites/lananhnguyen/2014/02/11/exclusive-flappy-bird-creator-dong-nguyen-says-app-gone-forever-because-it-was-an-addictive-product/?sh=c69e06a64760

41. Adam Alter, *Irresistible: The Rise of Addictive Technology and the Business of Keeping Us Hooked* (New York: Penguin Press, 2017) p.188.

42. Graeff, Erhardt, "The Responsibility to Not Design and the Need for Civic Professionalism," Dec. 16, 2020 on *BOW Big Ideas*, https://www.youtube.com/watch?v=f1qQCKGjpZg.

43. Canela Lopez, "7 Tech Executives Who Raise Their Kids Tech-Free or Seriously Limit Their Screen Time," March 4, 2020 on *BusinessInsider.com*, https://www.businessinsider.co.za/tech-execs-screen-time-children-bill-gates-steve-jobs-2019-9

44. This and the following Sacasas quotes are from an interview with the author on March 28, 2022

45. From Amish Studies: The Young Center, https://groups.etown.edu/amishstudies/cultural-practices/technology/

46. Ghosh, *Terms of Disservice*, p. 185, referencing Hunt Allcott, Luca Braghieri, Sarah Eichmeyer, Matthew Gentzkow, "The Welfare Effects of Social Media," NBER Working Paper No. 25514, Nov. 2019

47. https://www.statista.com/statistics/264810/number-of-monthly-active-facebook-users-worldwide/

48. Jeff Desjardins, "How Google Retains More Than 90 Percent of Market Share," April 23, 2018 on *BusinessInsider.com*, https://www.businessinsider.com/how-google-retains-more-than-90-of-market-share-2018-4

49. McNamee, *Zucked*, P. 334

50. Dahlia Lithwick and Mark Joseph Stern, "The Clarence Thomas Takeover," Aug. 2, 2017 on *Slate.com*, https://slate.com/news-and-politics/2017/08/clarence-thomas-legal-vision-is-becoming-a-trump-era-reality.html

51. Michael J. Wright, "The Decline of American Unions Is a Threat to Public Health," June 2016 in the *American Journal of Public Health*, https://www.ncbi.nlm.nih.gov/pmc/articles/PMC4880254/

52. Ted Van Green, "Majorities of Adults See Decline of Union Membership as Bad for the U.S. and Working People," Mar. 12, 2024 on *PewResearch.org*, https://www.pewresearch.org/short-reads/2024/03/12/majorities-of-adults-see-decline-of-union-membership-as-bad-for-the-us-and-working-people/

53. Jasmine Kerrissey and Judith Stepan-Norris, "This Labor Day Comes Amid the Biggest Jump in Union Activity in Decades," Sept. 2, 2022 in the *New York Times*, https://www.washingtonpost.com/politics/2022/09/02/labor-day-unions-workers-wages/

54. Sarah Kessler, Ephrat Livni, and Michael J. de la Merced, "Tech Fears Are Showing Up On Picket Lines," Sept. 16, 2023 on *NYTimes.com*, https://www.nytimes.com/2023/09/16/business/dealbook/uaw-strike-tech-ai-unions.html#:~:text=Unions%20aren%27t%20just%20fighting,this%20-gloomy%20outlook%20is%20inevitable.

55. Jonathan Vanian, "How Unions Are Pushing Back Against the Rise of Workplace Technology," Apr. 30, 2019 on *Fortune.com*, https://fortune.com/long form/unions-workplace-technology/

56. Dara Kerr, "Protestors, Armed with Traffic Cones, Are Immobilizing Driverless Cars," Aug. 28, 2023 on *NPR.org*, https://www.npr.org/2023/08/28/1196283050/protesters-armed-with-traffic-cones-are-immobilizing-driverless-cars#:~:text=The%20group%20figured%20out%20that,pig%20-for%20this%20new%20technology.

57. See the safety evidence in "ConeSF: a Campaign to Rein in Robotaxis," Oct. 24, 2023 on *SafeStreetRebel.com*, https://www.safestreetrebel.com/conesf/

58. Mumford, *Technics & Civilization*, pp. 284-285

59. Crawford, *Atlas of A.I.*, p. 86

60. Adam Seth Litwin, "Want to Save Your Job from A.I.? Hollywood Screenwriters Just Showed You How," Sept. 29, 2023 on *NYTimes.com*, https://www.nytimes.com/2023/09/29/opinion/wga-strike-deal-ai-jobs.html

61. "Judge Approves $1.5 Billion Settlement Over AI Company Anthropic's Alleged Use of Pirated Books," Sept. 25, 2025 on *CBSNewscom*, https://www.cbsnews.com/news/judge-approves-1-5-billion-dollar-settlement-anthropic-pirated-books/

62. Cecilia Kang, Ryan Mac, and Eli Tan, "Meta and YouTube Found Negligent in Landmark Social Media Addiction Case," March 25, 2026, on *NYTimes.com*, https://www.nytimes.com/2026/03/25/technology/social-media-trial-verdict.html

16. System Repairs

1. Robinson, *The Ministry for the Future*, pp. 251-252

2. McNamee, *Zucked*, p. 251

3. Liu, *Abolish Silicon Valley*, p. 46

4. Summer Moore Batte, "The Zoom Where It Happens," May 2021 on *StandfordMag.com*, https://stanfordmag.org/contents/the-zoom-where-it-happens

5. Phyllis Jordan, "Charting an Ethical Course for Technology's Future," Sept. 13, 2021 on *Future-Ed.org*, https://www.future-ed.org/charting-an-ethical-course-for-the-technology-future/

6. Oren Etzioni, "How to Regulate Artificial Intelligence," Sept. 1, 2017 in the *New York Times*, https: nytimes.com/2017/09/01/opinion/artificial-intelligence-regulations-rules.html

7. For the latter see, for example, Christine Seifert and Russell Clayton, "What Reading Fiction Can Teach Graduate Students About Empathy and Emotion," Feb. 22, 2021, https://hbsp.harvard.edu/inspiring-minds/what-reading-fiction-can-teach-graduate-students-about-empathy-and-emotion

8. Zuboff, *The Age of Surveillance Capitalism*, p. 291

9. Interview with the author, Feb. 7, 2022

10. See, for example, Jon Mertz, "Why You Should Consider Having Employees on Your Board," Dec. 9, 2019 on *Medium.com*, https://medium.com/swlh/why-you-should-consider-having-employees-on-your-board-6676d957aedf

11. https://en.wikipedia.org/wiki/Worker_representation_on_corporate_boards_of_directors

12. Press release: "Warren Joins Baldwin, Colleagues in Effort to Give Workers a Great Voice at Public Companies," Oct. 16, 2018, https://www.warren.senate.gov/oversight/letters/warren-joins-baldwin-colleagues-in-effort-to-give-workers-a-greater-voice-at-public-companies

13. Noah Kulwin, "Shoshana Zuboff on Surveillance Capitalism's Threat to Democracy" Feb. 24, 2019 in *New York*, https://nymag.com/intelligencer/2019/02/shoshana-zuboff-q-and-a-the-age-of-surveillance-capital.html

14. See, for example "Depression and Your Sense of Control" on *Clinical-Depression.co.uk*, https://www.clinical-depression.co.uk/dlp/understanding-depression/depression-and-your-sense-of-control/

15. See, for example, "User Centered Design," https://www.interaction-design.org/literature/topics/user-centered-design

16. Tarnoff & Weigel, *Voices from the Valley*, p. 33

17. https://www.apple.com/privacy/control/

18. McNamee, *Zucked*, p. 112

19. Ezra Klein, "What Biden's Top A.I. Thinker Concluded We Should Do," Tues. April 11, 2023 in the *New York Times*, https://www.nytimes.com/2023/04/11/opinion/ezra-klein-podcast-alondra-nelson.html?showTranscript=1

20. McNamee, ibid., pp. 112-113

21. Angus Hervey, "Move Slowly, and Don't Break Things," May 24, 2018 on *Medium.com*,https://medium.com/future-crunch/move-slowly-and-dont-break-things-693f00601b19

22. Press Release from Congresswoman Anna G. Eshoo, "Eshoo, Schakowsky, Booker Introduce Bill to Ban Surveillance Advertising," Jan. 18, 2022, https://eshoo.house.gov/media/press-releases/eshoo-schakowsky-booker-introduce-bill-ban-surveillance-advertising

23. https://www.congress.gov/bill/117th-congress/house-bill/8152/text

24. "S.1596-REAL Political Advertisements Act," https://www.congress.gov/bill/118th-congress/senate-bill/1596/text?s=1&r=5

25. Lindsey Graham and Elizabeth Warren, "Lindsey Graham and Elizabeth Warren: When It Comes to Big Tech, Enough Is Enough," July 27, 2023 on *NYTimes.com*,https://www.nytimes.com/2023/07/27/opinion/lindsey-gra ham-elizabeth-warren-big-tech-regulation.html

26. Aaron Sankin, "What Does Facebook Mean When It Says It Supports 'Internet Regulations'?" Sept. 16, 2021 on *TheMarkup.org*, https://the markup.org/the-breakdown/2021/09/16/what-does-facebook-mean-when-it-says-it-supports-internet-regulations

27. Adam Satariano, "E.U. Takes Aim at Big Tech's Power with Landmark Digital Act," March 24, 2022 in the *New York Times*, https://www.nytimes. com/2022/03/24/technology/eu-regulation-apple-meta-google.html

28. Adam Satariano, "Meta Fined \$1.3 Billion for Violating E.U. Data Privacy Rules," May 22, 2023 in the *New York Times*, https://www.nytimes.com/ 2023/05/22/business/meta-facebook-eu-privacy-fine.html#:~:text=Meta% 20on%20Monday%20was%20fined,European%20Union%20data%20protec tion%20rules.

29. Julia Angwin, "The Gatekeepers of Knowledge Don't Want Us to See What They Know," July 14, 2023 on *NYTimes.com*, https://www.nytimes.com/ 2023/07/14/opinion/big-tech-european-union-journalism.html

30. Adam Satariano, "Europeans Take a Major Step Toward Regulating A.I.: A Draft Law in the European Parliament Has Become the World's Most Far-Reaching Attempt to Address the Potentially Harmful Effects of Artificial Intelligence," June 14, 2023 on *NYTimes.com*, https://www.nytimes.com/ 2023/06/14/technology/europe-ai-regulation.html

31. "Age-Appropriate Design: A Code of Practices for Online Services," Information Commissioners Office, U.K., Sept 2, 2020, https://ico.org.uk/media/for-organisations/guide-to-data-protection/key-data-protection-themes/age-appro priate-design-a-code-of-practice-for-online-services-2-1.pdf

32. Steven Pearlstein, "Here's the Inside Story of How Congress Failed to Rein in Big Tech," July 6, 2023 in the *WashingtonPost.com*, https://www.washington post.com/opinions/2023/07/06/congress-facebook-google-amazon-apple-reg ulation-failure/

33. Cristiano Lima, "Lawmakers 'Cowered' to Take on Tech Giants Again, Senator Says," May 12, 2023 in the *Washington Post*, https://www.washing tonpost.com/politics/2023/05/12/lawmakers-cowered-take-tech-giants-again-senator-says/

34. Cristiano Lima, loc. cit.

35. Grant Gallagher, "Why AI Panic Is Not About Safety," June 5, 2023 on *Newsweek.com*, https://www.newsweek.com/why-ai-panic-not-about-safety-opinion-1804140

36. Cat Zakrzewski and Cristiano Lima, "Congress Is Racing to Regulate AI. Silicon Valley Is Eager to Teach Them How," June 17, 2023 on *Washington-Post.com*,https://www.washingtonpost.com/technology/2023/06/17/con gress-regulating-ai-schumer/

37. Cat Zakrzewski, "Washington Vows to Tackle AI, as Tech Titans and Critics Descend," Apr. 8, 2023 on *WashingtonPost.com*, https://www.washington post.com/technology/2023/04/08/washington-artificial-intelligence-regula tion/

38. See, for example, Cecilia Kang, "In U.S., Regulating A.I. Is in Its 'Early Days'," July 21, 20123 on *NYTimes.com*, https://www.nytimes.com/2023/ 07/21/technology/ai-united-states-regulation.html; and Kevin Roose, "How Do the White House's A.I. Commitments Stack Up?" July 22, 2023 on *NYTimes.com*, https://www.nytimes.com/2023/07/22/technology/ai-regula tion-white-house.html

39. Interview with the author, April 10, 2023

40. "Remarks by Vice President Vance at the Artificial Intelligence Action Summit," Feb. 11, 2025 on *WhiteHouse.gov*, https://www.whitehouse.gov/ remarks/2025/02/remarks-by-vice-president-vance-at-the-artificial-intelli gence-action-summit/

41. David McCabe, "Google Broke the Law to Keep Its Advertising Monopoly, a Judge Rules," April 17, 2025 on *NYTimes.com*, https://www.nytimes.com/ 2025/04/17/technology/google-ad-tech-antitrust-ruling.html

42. John Herrman, "The New Green Light for Google's Monopoly," Sept. 9, 2025 on *NYMag.com*, https://nymag.com/intelligencer/article/googles-monopoly-isnt-going-anywhere.html

43. "Governor Newsom Signs Data Privacy Bills to Protect Tech Users," Oct. 8, 2025 press release from Governor Gavin Newsom, https://www.gov.ca.gov/ 2025/10/08/governor-newsom-signs-data-privacy-bills-to-protect-tech-users/

17. A Cornucopia Of Other Solutions

1. Quoted in "How to Fix Social Media," by various authors in Oct. 29, 2021 in the *Wall Street Journal*, https://www.wsj.com/tech/how-to-fix-social-media-11635526928#jaron-lanier-topple-the-new-gods-32eb0f70

2. Quoted in How to Fix Social Media," ibid.

3. Gavet, *Trampled by Unicorns*, pp. 177-178

4. Gerrit De Vynck, "AI Images Are Getting Harder to Spot. Google Thinks It Has a Solution," Aug. 29, 2023 on *WashingtonPost.com*, https://www.wash ingtonpost.com/technology/2023/08/29/google-wants-watermark-ai-gener ated-images-stop-deepfakes/

5. Roddy Lindsay, "I Designed Algorithms at Facebook. Here's How to Regulate Them," Oct. 6, 2021in the *New York Times*, https://www.nytimes.com/ 2021/10/06/opinion/facebook-whistleblower-section-230.html

6. Aviv Ovadya and Luke Thorburn, "Bridging Systems: Open Problems for Countering Destructive Division Across Ranking, Recommenders, and Governance," July 24, 2023, draft of a study to be published by the Knight

Notes

First Amendment Institute at Columbia University, https://arxiv.org/pdf/2301.09976.pdf

7. Ghosh, *Terms of Disservice*, pp. 230-231
8. Hanna Rosin, "The Smartphone Kids Are Not All Right," Mar. 21, 2024 on *TheAtlantic.com*, https://www.theatlantic.com/podcasts/archive/2024/03/smartphone-anxious-generation-mental-health/677817/
9. Haidt quoted in Rosin, ibid.
10. Holly Yan and Debra Goldschmidt, "Los Angeles School Board Will Ban Students from Using Cell Phones During the School Day. But Questions Loom About How to Do It," June 18, 2024 on *CNN.com*, https://www.cnn.com/2024/06/18/us/lausd-cell-phone-ban-students/index.html
11. Natasha Singer, "Why Schools Are Racing to Ban Student Phones," Aug. 11, 2024 on *NYTimes.com*, https://www.nytimes.com/2024/08/11/technology/school-phone-bans-indiana-louisiana.html
12. "Distraction-Free Schools: Governor Hochul Announces New York to Become Largest State in the Nation with Statewide, Bell-to-Bell Restrictions on Smartphones in Schools," May 6, 2025 on *Governor.ny.gov*, https://www.governor.ny.gov/news/distraction-free-schools-governor-hochul-announces-new-york-become-largest-state-nation
13. Jonathan Haidt, Will Johnson, and Zach Rausch, "The Smartphones Haven't Defeated Us. Yet," June 18, 2025 on *NYTimes.com*, https://www.nytimes.com/2025/06/18/opinion/parents-smartphones-tiktok-facebook.html
14. McNamee, *Zucked*, p. 321
15. McNamee, ibid., p. 323
16. McNamee, ibid., pp. 283-284
17. McNamee, ibid., p. 323
18. Gavet, ibid., p. 156
19. Taplin, *Move Fast and Break Things*, p. 282
20. See, for example, John Markoff, "A Long-Shot Bet to Bypass the Middlemen of Social Media," March 25 2025 on *NYTimes.com*, https://www.nytimes.com/2025/03/06/technology/mike-mccue-surf-browser-decentralized-internet.html; "What You Need to Know About Decentralized Social Networks," from Tulane University School of Professional Advancement, https://sopa.tulane.edu/blog/decentralized-social-networks; and Justin Pot, "How the 'Fediverse' Works (and Why It May Be the Future of Social Media)," Oct. 25, 2024 on *Lifehacker.com*, https://lifehacker.com/tech/what-is-the-fediverse-the-potential-future-of-social-media.
21. Aram Sinreich and Robert W. Gehl, "The Fediverse Promises Social Media Without Big Tech—If It Can Avoid Familiar Pitfalls," Mar. 12, 2025 on *The Conversation.com*, https://theconversation.com/the-fediverse-promises-social-media-without-big-tech-if-it-can-avoid-familiar-pitfalls-247178
22. Doctorow, *How to Destroy Surveillance Capitalism*, p. 125
23. Wu, *The Master Switch*, p. 304

24. Patrick Kingsley, "The Wu Master," March 17, 2011 on *TheGuardian.com*, https://www.theguardian.com/technology/2011/mar/17/the-master-switch-tim-wu-internet

25. McNamee, ibid., p. 300

26. See, for example, Dana Mattioli, "Amazon Scooped Up Data from Its Own Sellers to Launch Competing Products," April 23, 2020 in the *Wall Street Journal*,https://www.wsj.com/articles/amazon-scooped-up-data-from-its-own-sellers-to-launch-competing-products-11587650015

27. Gavet, ibid., pp. 134-135

28. Gavet, ibid., p. 138

29. Doctorow, ibid., p. 124

30. Doctorow, ibid., p. 123

31. Bill Joy, "Why the Future Doesn't Need Us," April 1, 2000 in *Wired*, https://www.wired.com/2000/04/joy-2/

32. Reich, Sahami, & Weinstein, *System Error*, p. xxix

33. Reich, Sahami, & Weinstein, ibid., p. 245

34. Will Oremus, "Facebook Keeps Researching Its Own Harms—And Burying the Findings," Sept. 16, 2021 in the *Washington Post*, https://www.washingtonpost.com/technology/2021/09/16/facebook-files-internal-research-harms/

35. Kevin Roose, "Facebook's Frankenstein Moment," Sept. 21, 2017 in the *New York Times*, https://nytimes.com/2017/09/21/technology/facebook-frankenstein-sandberg-ads.html

36. "Putting Principles Into Practice at Microsoft," https://www.microsoft.com/en-us/ai/principles-and-approach

37. Kevin Roose, "Bing's A.I. Chat: 'I Want to Be Alive,'" Feb. 16, 2003 in the *New York Times*, https://www.nytimes.com/2023/02/16/technology/bing-chatbot-transcript.html

38. See, for example, Zeynep Tufekci, "Why Zuckerberg's 14-Year Apology Tour Hasn't Fixed Facebook," Apr. 6, 2018 in *Wired*, https://www.wired.com/story/why-zuckerberg-15-year-apology-tour-hasnt-fixed-facebook/

39. Reich, Sahami, & Weinstein, ibid., p. 150

40. Scoblic, ibid.

41. For more on this, see Steven Johnson, "Magic Cookies," July 13, 2020 on Johnson's *Adjacent Possible* blog, https://adjacentpossible.substack.com

42. The above quotes are from an interview with the author on March 11, 2022

43. Franklin, *The Real World of Technology*, p. 65

44. Wallach, *A Dangerous Master*, p. 250

45. Gary Marcus, quoted in David Marchese, "How Do We Ensure an A.I. Future That Allows for Human Thriving?" May 1, 2023 in the *New York Times Magazine*, https://www.nytimes.com/interactive/2023/05/02/magazine/07mag-talk-marcus.html

46. Wallach, ibid., p. 255

47. Wallach, ibid., pp. 265 and 282

48. Interview with the author, ibid.

49. Interview with the author, ibid.

50. Michael Newell, "Top 5 Largest Fines Levied on Tech Companies by the European Commission," Feb. 18, 2019 in *TheNewEconomy.com*, https://www.theneweconomy.com/business/top-5-largest-fines-levied-on-tech-companies-by-the-european-commission

51. Interviewed by Kara Swisher, "Brazen Is the Order of the Day," Oct. 5, 2021 in the *New York Times*, https://www.nytimes.com/2021/10/05/opinion/facebook-blackout-2021.html

52. Gary Marcus, quoted in David Marchese, "How Do We Ensure an A.I. Future That Allows for Human Thriving?" May 1, 2023 in the *New York Times Magazine*, https://www.nytimes.com/interactive/2023/05/02/magazine/07mag-talk-marcus.html

53. See, for example, Mark McCarthy, "Transparency Is the Best First Step Towards Better Digital Governance," May 10, 2022 on *Brookings.edu*, https://www.brookings.edu/articles/transparency-is-the-best-first-step-towards-better-digital-governance/

54. Interview with the author, Mar. 25, 2022

55. Kashmir Hill and Daisuke Wakabayashi, "Google Seeks to Break Vicious Cycle of Online Slander," June 10, 2021 in the *New York Times*, https://www.nytimes.com/2021/06/10/technology/google-algorithm-known-victims.html?action=click&module=In%20Other%20News&pgtype=Homepage

56. Quoted anonymously in Tarnoff & Weigel, *Voices from the Valley*, pp. 103-104

57. Mathew Ingram, "Is Facebook Quitting the News Business?" Dec. 7, 2022 in the *Columbia Journalism Review*, https://www.cjr.org/cjr_outbox/is-facebook-quitting-the-news-business.php

58. Deudney, *Dark Skies*, p. 139

59. Deudney, ibid., p. 275

60. McNamee, *Zucked.*, p. 297

61. Above quotes from an interview with the author, Feb. 4, 2022

62. Quoted in Deudney, ibid., p. 225

63. United Nations Office for Outer Space Affairs, "Treaty on Principles Governing the Activities of States in the Exploration and Use of Outer Space, including the Moon and Other Celestial Bodies," https://www.unoosa.org/oosa/en/ourwork/spacelaw/treaties/introouterspacetreaty.html

64. Jocelyn Kaiser, "NIH Lifts 3-Year Ban on Funding Risky Virus Studies," Dec. 19, 2017 on *Science.org*, https://www.science.org/content/article/nih-lifts-3-year-ban-funding-risky-virus-studies

65. Peter Shanks, "Drawing a Line: European Convention Upholds Ban on Heritable Human Genome Editing," Nov. 3, 2022 on *GeneticLiteracyProject.org*, https://geneticliteracyproject.org/2022/11/03/drawing-a-line-european-convention-upholds-ban-on-heritable-human-genome-editing/

66. For a list of these countries, see "Human Cloning" on *Wikipedia.org*, https://en.wikipedia.org/wiki/Human_cloning

67. Bill McKibben, "Dimming the Sun to Cool the Planet Is a Desperate Idea, Yet We're Inching Toward It," Nov. 22, 2022 in the *New Yorker*, https://www.newyorker.com/news/annals-of-a-warming-planet/dimming-the-sun-to-cool-the-planet-is-a-desperate-idea-yet-were-inching-toward-it

68. Secretariat of the Convention on Biological Diversity, "Geoengineering in Relation to the Convention on Biological Diversity: Technical and Regulatory Matters," Sept. 12, on *CBD.int*, https://www.cbd.int/doc/publications/cbd-ts-66-en.pdf

69. "Open Letter: We Call for an International Non-Use Agreement on Solar Geoengineering,"https://www.solargeoeng.org/non-use-agreement/open-letter/

70. Justine Calma, "Mexico Bans Solar Geoengineering Experiments After Start-up's Field Tests," Jan. 18, 2023 on *TheVerge.com*, https://www.theverge.com/2023/1/18/23560446/mexico-ban-solar-geoengineering-make-sunsets-startup-experiments

71. Eliezer Yudkowsky, "Pausing AI Developments Isn't Enough. We Need to Shut It All Down," Mar. 29, 2023 on *Time.com*, https://time.com/6266923/ai-eliezer-yudkowsky-open-letter-not-enough/#:~:text=If%20we%20held%20anything%20in,in%20any%20reasonable%20time%20window

72. "Pause Giant AI Experiments: An Open Letter," March 22, 2023 on *FutureOfLife.org*, https://futureoflife.org/open-letter/pause-giant-ai-experiments/#:~:text=We%20call%20on%20all%20AI

73. Yudkowsky, ibid.

74. Casey Newton and Kevin Roose, "All Gas, No Brakes in A.I.+Metaverse Update+Lessons from a Prompt Engineer," Sept. 29, 2023 on *NYTimes.com*, https://www.nytimes.com/2023/09/29/podcasts/all-gas-no-brakes-in-ai-metaverse-update-lessons-from-a-prompt-engineer.html?showTranscript=1

18. Other Ways To Go

1. Mumford, *Technics & Civilization*, p. 281

2. Quoted in Derrick O'Keefe, "Imagining the End of Capitalism with Kim Stanley Robinson," Oct. 22, 2020 on *Jacobin.com*, https://jacobin.com/2020/10/kim-stanley-robinson-ministry-future-science-fiction; actually, this seems to be a paraphrase of Jameson, via Slavoj Zizek, and then Mark Fisher, according to Robert Scott, "Postcritique; or, the Cultural Logic of Capitalist Realism," Sept. 14, 2022 in *LaReviewofBooks.com*, lareviewofbooks.org/article/postcritique-or-the-cultural-logic-of-capitalist-realism/

3. Shoshana Zuboff "You Are the Object of a Secret Extraction Operation" Nov. 11, 2021 in the *New York Times*, https://www.nytimes.com/2021/11/12/opinion/facebook-privacy.html

4. Ghosh, *Terms of Disservice.*, p. 242

5. Ben Tarnoff, "The Internet Is Broken. How Do We Fix It?" May 27, 2022 in the *New York Times*, https://www.nytimes.com/2022/05/27/opinion/technology/what-would-an-egalitarian-internet-actually-look-like.html?action=click&module=Well&pgtype=Homepage§ion=Guest%20Essays

6. O'Keefe, ibid.

7. "Reality is an Activity of the Most August Imagination: A Conversation with Tim O'Reilly," Oct. 2, 2017 on *Edge.org*, https://www.edge.org/conversation/tim_oreilly-reality-is-an-activity-of-the-most-august-imagination

8. L.M. Sacasas, "Note from the Metaverse," Sept. 4, 2021 on *The Convivial Society*, https://theconvivialsociety.substack.com/p/notes-from-the-metaverse

9. See, for example, Zia Qureshi, "Tackling the Inequality Pandemic: Is There a Cure?" Nov. 17, 2020 on the *Brooking Institution* website, https://www.brookings.edu/research/tackling-the-inequality-pandemic-is-there-a-cure/

10. Nick Drozd comment on "Why Am I Not Terrified of AI?" Mar. 6, 2023 on *Shtetl Optimized: The Blog of Scott Aaronson*, https://scottaaronson.blog/?p=7064

11. From Ted Chiang, "Will A.I. Become the New McKinsey?" May 4, 2023 in the *New Yorker*, https://www.newyorker.com/science/annals-of-artificial-intelligence/will-ai-become-the-new-mckinsey

12. Chiang, ibid.

13. https://www.symbotic.com; see also https://www.marineinsight.com/know-more/what-are-dark-warehouses/

14. Mumford, ibid., p. 402

15. Mark O'Connell, "Why Silicon Valley Billionaires Are Prepping for the Apocalypse in New Zealand," Feb. 15, 2018 on the *Guardian.com*, https://www.theguardian.com/news/2018/feb/15/why-silicon-valley-billionaires-are-prepping-for-the-apocalypse-in-new-zealand

16. Guthrie Scrimgeour, "Inside Mark Zuckerberg's Top-Secret Hawaii Compound," Dec. 14, 2023 on *Wired.com*, https://www.wired.com/story/mark-zuckerberg-inside-hawaii-compound/

17. Noah Kulwin, "Shoshana Zuboff on Surveillance Capitalism's Threat to Democracy" Feb. 24, 2019 in *New York*, https://nymag.com/intelligencer/2019/02/shoshana-zuboff-q-and-a-the-age-of-surveillance-capital.html

18. McNamee, *Zucked*, p. 236

19. From Noah Kulwin, "The Internet Apologizes," Apr. 16, 2008 in *New York*, https://nymag.com/intelligencer/2018/04/an-apology-for-the-internet-from-the-people-who-built-it.html

20. Interview with the author, March 13, 2022

21. Interview with the author, March 13, 2022

22. "How We Invest," https://www.kaporcapital.com/how-we-invest/

23. Curt Nickish, "The VC Fund Closing Equity Gaps—and Making Money," Aug. 8, 2023 on the *Harvard Business Review HBR IdeaCast* podcast, https://

hbr.org/podcast/2023/08/the-vc-fund-closing-equity-gaps-and-making-money

24. Nickish, loc. cit.

25. Interview with the author, March 13, 2022

26. Rushkoff, *Throwing Rocks at the Google Bus*, p. 11

27. Rushkoff, ibid., p. 95

28. Rushkoff, ibid., p. 135

29. Rushkoff, ibid., p. 119

30. Ezra Klein, "Sam Altman on the A.I. Revolution, Trillionaires and the Future of Political Power," June 11, 2021 in the *New York Times*, https://www.nytimes.com/2021/06/11/opinion/ezra-klein-podcast-sam-altman.html?action=click&module=Opinion&pgtype=Homepage&showTranscript=1

31. Interview with the author, Mar. 11, 2022

32. Liz Hoffman and Reed Albergotti, "Microsoft Eyes $10 Billion Bet on Chat-GPT," Jan. 9, 2023 on *Semafor.com*, https://www.semafor.com/article/01/09/2023/microsoft-eyes-10-billion-bet-on-chatgpt

33. Kevin Roose, "How ChatGPT Kicked Off an A.I. Arms Race," Feb. 3, 2023 in the *New York Times*, https://www.nytimes.com/2023/02/03/technology/chatgpt-openai-artificial-intelligence.html

34. See, for example, tech journalists Kevin Roose and Casey Newton speaking on the Ezra Klein Show podcast on *NYTimes.com*, https://www.nytimes.com/2023/12/01/opinion/ezra-klein-podcast-casey-newton-kevin-roose.html?showTranscript=1

35. Will Knight, "OpenAI's Long-Term Risk Team Has Disbanded," May 17, 2024 on *Wired.com*, https://www.wired.com/story/openai-superalignment-team-disbanded/

36. Interview with the author, March 25, 2022

37. Liu, *Abolish Silicon Valley*, pp. 190-191

38. Lanier, *You Are Not a* Gadget, pp. 106-107

39. Robinson, *Ministry for the* Future, ibid. p. 281

40. Interview with the author, March 13, 2022

41. Gavet, *Trampled by Unicorns*, p. 157

42. Interview with the author, March 28, 2022

43. Klein and Olson, ibid.

44. Robinson, ibid., p. 294

45. Robinson, ibid., 454

46. Interview with the author, March 13, 2022

47. Charles Harvey & Kurt House, "Every Dollar Spent on This Climate Technology Is a Waste," Aug. 16, 2022 in the *New York Times*, https://www.nytimes.com/2022/08/16/opinion/climate-inflation-reduction-act.html

48. Liu, ibid., p. 137

49. Rushkoff, ibid., p. 228

50. Rushkoff, ibid., pp. 227-228

51. Rushkoff, ibid., p. 231

52. Rushkoff, ibid., pp. 224-225

53. Nick Romeo, "How Mondragon Became the World's Largest Co-op," Aug. 27, 2022 in the *New Yorker*,https://www.newyorker.com/business/currency/how-mondragon-became-the-worlds-largest-co-op

54. Robinson describes the real-world co-op in his novel *The Ministry for the Future*, pp. 271-273

55. Romeo, ibid.

56. Romeo, ibid.

57. O'Reilly, ibid.

58. Tarnoff & Weigel, *Voices from the Valley*, pp. 33-34

59. Liu, ibid., p. 197

60. Ben Tarnoff, "The Internet Is Broken. How Do We Fix It?" May 27, 2022 in the *New York Times*, https://www.nytimes.com/2022/05/27/opinion/technology/what-would-an-egalitarian-internet-actually-look-like.html?action=click&module=Well&pgtype=Homepage§ion=Guest%20Essays

61. Tarnoff, loc. cit.

62. Tarnoff, loc. cit.

63. Larry Elliott, "World's Eight Richest People Have Same Wealth as Poorest 50%," Jan. 15, 2017 on *TheGuardian.com*, https://www.theguardian.com/global-development/2017/jan/16/worlds-eight-richest-people-have-same-wealth-as-poorest-50

64. Oxfam International press release, "Ten Richest Men Double Their Fortunes in Pandemic While Incomes of 99 Percent of Humanity Fall," Jan. 17, 2022, https://www.oxfam.org/en/press-releases/ten-richest-men-double-their-fortunes-pandemic-while-incomes-99-percent-humanity

19. Rising To The Challenge

1. Liu, *Abolish Silicon Valley*, p. 207

2. Interview by Andrew Anthony, "We Were Incentivized to Have a Bad Pandemic Response," May 9, 2021 on the *Guardian.com*. https://www.theguardian.com/books/2021/may/09/michael-lewis-the-premonition-covid-response-the-big-short-interview

3. L.M. Sacasas, "The Tech Backlash We Really Need," Spring 2018 in *The New Atlantis*, https://www.thenewatlantis.com/publications/the-tech-backlash-we-really-need

4. Derek Thompson, "Why Are American Teenagers So Sad and Anxious," Apr. 22, 2022 in *TheRinger.com*, theringer.com/2022/4/22/23036468/why-are-american-teenagers-so-sad-and-anxious

5. Interview with the author, Mar. 28, 2022

6. Interview with the author, Mar. 28, 2022

7. Jaron Lanier, "How to Fix Social Media," Oct. 29, 2021 in the *Wall Street*

*Journal,*https://www.wsj.com/articles/how-to-fix-social-media-11635526928#jaron-lanier-topple-the-new-gods-32eb0f70

8. L.M. Sacasas, "The Tech Backlash We Really Need," Spring 2018 in *The New Atlantis,* https://www.thenewatlantis.com/publications/the-tech-back lash-we-really-need

9. Hall, *Where Is My Flying Car?*, pp. 28-281

10. Hall, ibid., p. 287 (I'm citing just one instance, but this is one of Hall's favorite epithets for anyone who would dare to critique technologies, and he lards it throughout the book.)

11. Hall, ibid., p. 286

12. Hall, ibid., p. 284

13. Hall, ibid., p. 290

14. These quotes are from an interview with the author on Aug. 23, 2022

15. "Reality is an Activity of the Most August Imagination: A Conversation With Tim O'Reilly," Oct. 2, 2017 on Edge.org, https://www.edge.org/conversa tion/tim_oreilly-reality-is-an-activity-of-the-most-august-imagination

16. Borsook, *Cyberselfish*, p. 112

17. For an interesting discussion about attention, see "Transcript: Ezra Klein Interviews D. Graham Burnett," May 31 on *NYTimes.com*, https://www. nytimes.com/2024/05/31/podcasts/transcript-ezra-klein-interviews-d-gra ham-burnett.html

18. Alex Vadukul, "Luddite Teens Don't Want Your Likes," Dec. 15, 2022 on *NYTimes.com,*https://www.nytimes.com/2022/12/15/style/teens-social-media.html

19. Graeber & Wengrow, *The Dawn of Everything*, pp. 498-499